ELECTRICAL MOTOR CONTROLS
Automated Industrial Systems
WORKBOOK

Third Edition

AMERICAN TECHNICAL PUBLISHERS, INC.
HOMEWOOD, ILLINOIS 60430

Gary Rockis
Glen Mazur

3 4 5 6 7 8 9 - 92 - 9 8 7 6 5 4 3 2 1

Printed in the United States of America

ISBN 0-8269-1667-8

INTRODUCTION

This workbook is designed to let you practice what you have studied in the textbook, *Electrical Motor Controls, Automated Industrial Systems.* A quiz (Tech-Chek) and several worksheets are provided for each of the units in the textbook. This workbook also provides data sheets showing detailed technical procedures for solving common control problems, such as reversing a motor or applying time control.

The worksheets proceed step by step through the basic kinds of control circuits you will encounter on the job. Some worksheets will refer you to a data sheet, which you will find in the back of the workbook. When filling in the worksheets, use standard lettering, numbering, and coding information as detailed in the textbook, *Electrical Motor Controls, Automated Industrial Systems.* You can use these completed worksheets to wire actual electrical circuits on the job, so be sure to meet all industrial standards in completing them.

Acknowledgments

The author and publisher are grateful to the following companies who provided drawings and information for this workbook.

Eaton Corp. Cutler-Hammer Products
Electromatic Controls Corp.

PROGRESS SHEET

Complete the unit quiz and worksheets for each of the seventeen units in the textbook, *Electrical Motor Controls, Automated Industrial Systems.*

Date Completed

UNIT

1 Electrical Tools, Instruments and Safety _____

2 Industrial Electrical Symbols and Line Diagrams _____

3 Introduction to Logic as Applied to Line Diagrams _____

4 AC Manual Contactors and Motor Starters _____

5 Magnetism and Magnetic Solenoids _____

6 AC/DC Contactors and Magnetic Motor Starters _____

7 Time Delay and Logic Applied to More Complex Line Diagrams and Control Circuits _____

8 Application and Installation of Control Devices _____

9 Reversing Circuits Applied to Single-Phase, Three-Phase and DC Motor Types. _____

10 Power Distribution Systems, Transformers, Switchboards, Panelboards, Motor Control Centers and Busways _____

11 Solid State Electronic Control Devices _____

12 Electromechanical and Solid State Relays _____

13 Photoelectric and Proximity Control and Applications _____

14 Programmable Controllers _____

15 AC Reduced Voltage Starters _____

16 Accelerating and Decelerating Methods and Circuits _____

17 Preventive Maintenance and Troubleshooting _____

WORKSHEET 1-1

Name _____ Class _____ Date _____

Select the best way to complete each statement and circle a, b, or c. Do not circle more than one choice for each statement.

1. The best system for organizing tools to be used either on the job or at a test bench is the
    a. pegboard.
    b. electrician's pouch.
    c. portable tool box.

2. Cutting tools should be
    a. stainless steel.
    b. new.
    c. sharp and clean.

3. All power tools should be grounded unless
    a. you have permission to use them without grounding.
    b. the area is dry.
    c. they are double-insulated.

4. A change in sound during tool operation usually indicates
    a. normal operation.
    b. trouble of some type.
    c. the tool is in reverse.

5. When replacing fuses, you should install a fuse
    a. first into the line side, then the load side.
    b. first into the load side, then the line side.
    c. into the load and line side simultaneously.

6. Rags containing oil, gasoline, alcohol, shellac, paints, varnish, or lacquer must be
    a. kept in a covered metal container.
    b. stored in a wastebasket.
    c. stored in a cool, dry place.

7. All unfamiliar wires should be treated as if they were
    a. dead.
    b. alive.
    c. harmless.

8. Acid on hands and face should be immediately washed away with plenty of
    a. water.
    b. glycerine.
    c. vaseline.

9. Shorting wires together is not safe because it
    a. makes the current close.
    b. may start a fire.
    c. may burn a fuse.

WORKSHEET 1-2

Score _____

Name _____ Class _____ Date _____

Match the items on the left with the answers on the right by entering the appropriate letter from the right in each blank space. Enter only one letter per space. Do not use a letter more than once.

_____ 1. Color meaning exposed parts of pulleys, gears, and rollers

_____ 2. Color meaning radiation hazard

_____ 3. Color meaning fire equipment

_____ 4. Color meaning caution

_____ 5. Color meaning first aid equipment

_____ 6. Used to secure wire harnesses

_____ 7. Used to remove insulation from small-diameter wires

_____ 8. Used in isolating control circuit components

_____ 9. Used for testing insulation

_____ 10. Used to monitor voltage and current variations

_____ 11. Used to bend wire, cut wire, and position small components.

_____ 12. Used to provide threading for control panel mounting

_____ 13. Used to check for defective controls with power off

_____ 14. Used to deburr conduit

_____ 15. Used to determine proper phasing

_____ 16. Used to identify wires

_____ 17. Used to secure fasteners in concrete

_____ 18. Used to cut cable and remove knockouts

_____ 19. Used in leveling long conduit runs

_____ 20. Warns that equipment is being worked on

_____ 21. Used to punch holes in metal

_____ 22. Used to pull wires through conduit

_____ 23. Used to verify grounding

_____ 24. Used to secure large conduits

_____ 25. Used to remove insulation from heavy duty cable

a. Red

b. Yellow

c. Orange

d. Purple

e. Green

f. Long needle-nosed plier

g. Lineman's side-cutting pliers

h. Cable strippers

i. Wire strippers

j. 24-inch level

k. Fish tape

l. Reaming tool

m. Hydraulic punch set

n. Power stud gun

o. Tie-rap gun

p. Lock-out tag

q. Polarized receptacle tester

r. Megohmmeter

s. Phase sequence indicator

t. Chain pipe wrench

u. Tap and die set

v. Wire markers

w. Continuity test

x. Strip chart recorder

y. Test lead set

TECH-CHEK ✔ 1

Name _____

Class _____ Date _____

Electrical Tools, Instruments and Safety

Score _____

Fill in the blanks to complete each statement.

1-3. Three systems for organizing electrical tools are the (1)_____,

(2)_____ _____, and (3)_____

_____ _____.

4. All power tools should be properly _____ before being used.

5. Before connecting a tool to a power source, you should make sure the switch is in the

_____ position.

6. When removing a fuse from the circuit, you should remove the _____ side

of the fuse first.

7. Before performing any repair on a piece of electrical equipment, you should be absolutely certain

the source of electricity is open and _____.

8. The color used to designate dangerous parts of machines and exposed parts

is _____.

9. The color used to designate fire protection equipment and emergency stop buttons and switches

is _____.

10. The color used to designate radiation hazards is _____.

11. The color used to designate caution is _____.

12. The color used to designate safety and the location of first aid equipment is

_____.

13. Fires occurring in wood, clothing, or paper are classified as Class _____.

14. Fires occurring in electrical equipment are classified as Class _____.

15. Fires occurring in flammable liquids, such as gasoline and oil, are classified as

Class _____.

16. Fires occurring in combustible metals, such as magnesium, are classified as

Class _____.

17. When checking for voltage in a circuit, the electrician should use a _____.

18. When checking a current level in a circuit, the electrician should use an

_____.

Continued

19. When checking for a resistance in a component or device (with power removed), the electrician should use an _____.

20. Before working on any electricial circuit, the number one rule an electrician should follow is: Never assume the power is _____.

WORKSHEET 2-1

Name _____ Class _____ Date _____

Using a ruler, draw the appropriate symbol below its description. Your lines should be straight and neatly drawn.

1. NO (normally open) limit switch

4. NO and NC pushbutton

2. Circuit breaker with thermal and magnetic overload

5. Single voltage transformer

3. NO held-closed limit switch

6. Dual voltage transformer

Continued

5

7. Three-phase motor

11. Red pilot light

8. NO timed closed contact

12. Photosensitive cell

9. Disconnect

13. Full wave rectifier

10. Two-position selector switch

14. Shunt field

6

WORKSHEET 2-2

Score _____

Name _____ Class _____ Date _____

Using a ruler, complete the schematic diagram with the appropriate symbol as called for in each of the circuit descriptions given below.

Circuit 1: NC held-open limit switch.

Circuit 2: NC timed closed contact.

Circuit 3: NC mushroom head pushbutton.

Circuit 4: NO temperature-activated switch.

Circuit 5: NO solid state limit switch.

Circuit 6: NC flow switch.

Circuit 7: Thermal overload device.

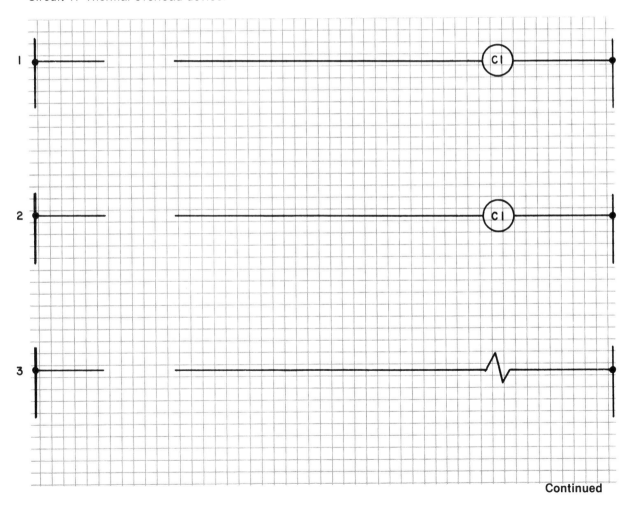

Continued

7

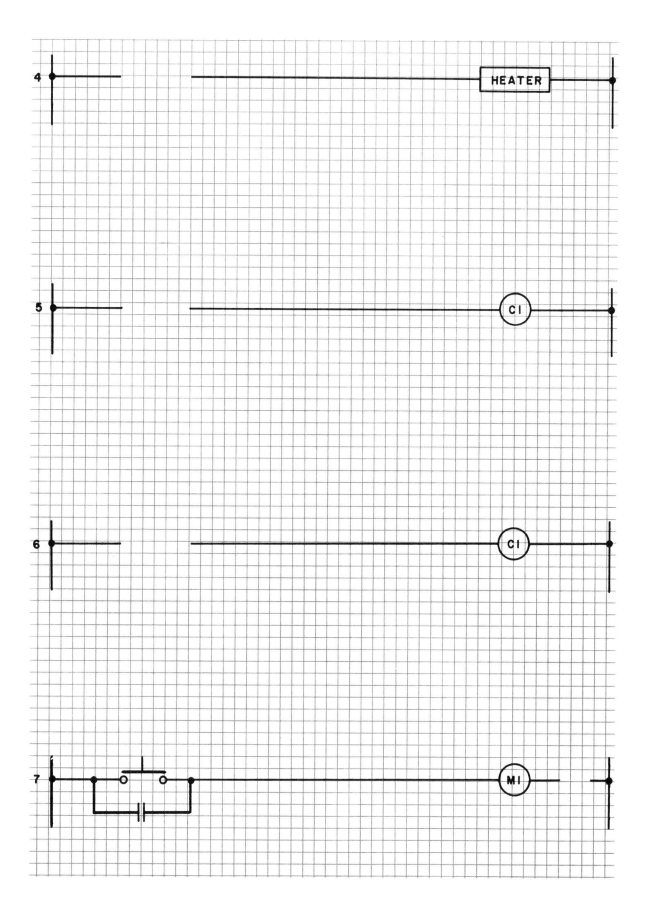

WORKSHEET 2-3

Score _____

Name _____ Class _____ Date _____

Complete each line diagram according to the circuit information given below. Use standard lettering, numbering, and coding information. Connecting lines should be straight and the circuits neatly drawn.

Circuit 1: A NO start pushbutton is to control a magnetic motor starter with three overload contacts.

Circuit 2: Redraw Circuit 1, adding an auxiliary contact to form a memory circuit. Add a NC stop pushbutton to turn the circuit off.

Circuit 3: Redraw Circuit 2, adding a foot switch and a limit switch that could also turn off the motor if actuated.

Circuit 4: Draw a control circuit in which a pressure switch is used to control a pump motor. The pump motor should turn on any time the pressure drops below 30 psi.

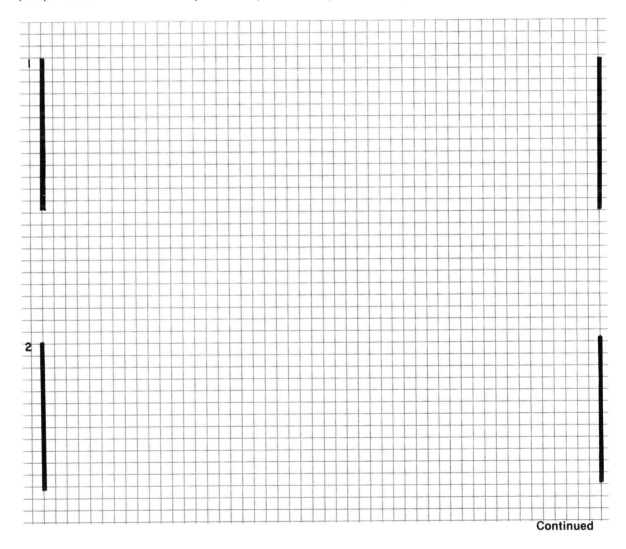

Continued

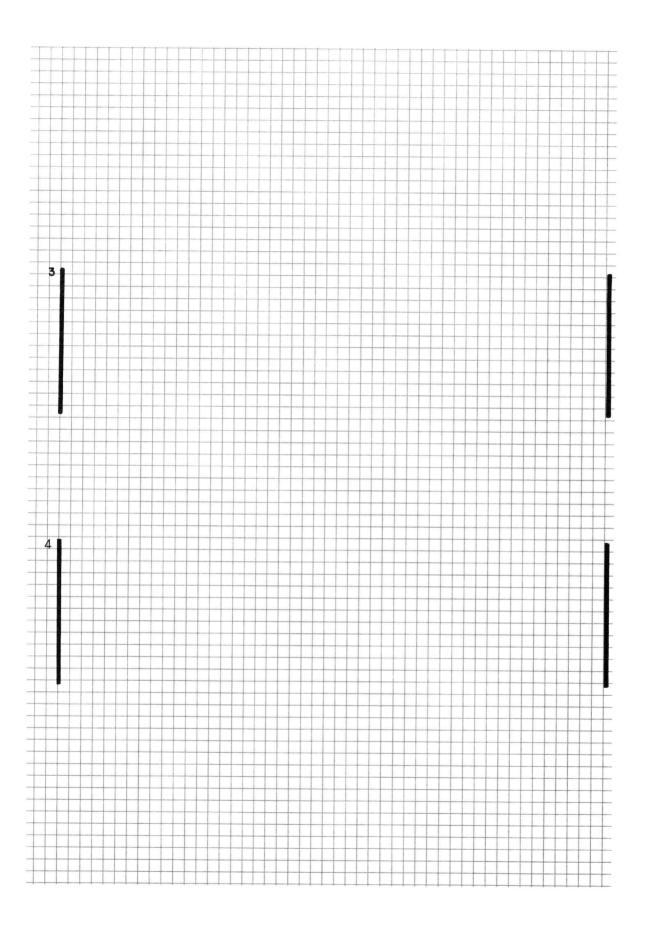

 **2**

Industrial Electrical Symbols and Line Diagrams

Fill in the blanks to complete each statement.

1. In electrical circuits, the basic means of communicating the language of control is the

   _____ diagram.

2. In a line diagram the power source is shown in _____ (thinner or heavier)

   lines than the rest of the diagram.

3. In a line diagram the path of current flow through the various parts of the control circuits, such as

   the pushbuttons, etc., is shown in _____ (thinner or heavier) lines than the

   rest of the diagram.

4. A pushbutton is an example of a _____ (manual or automatic) control

   switch.

5. A liquid level switch is an example of a _____ (manual or automatic) control

   switch.

6. A line diagram is always read from line _____ to

   line _____.

7. In a line diagram one side of the overload contact is connected to line

   _____.

8. When using a float switch to maintain a predetermined level, you should use the

   _____ contacts.

9. When using a float switch to control a sump pump, you should use the

   _____ contacts.

10. An electrical device which consists of a frame, plunger, and coil, and is used to create a push or

    pull type action, is the _____.

11. An electrical device which consists of a frame, plunger, and coil, and is used to open and close a

    set of contacts, is called a _____.

12. An electrical device which consists of a frame, plunger, and coil, and is used to open and close a

    set of contacts in addition to providing overload protection, is called a

    _____ _____ _____.

Continued

11

13. The _____ contacts are used in the control circuit to maintain an electrical holding circuit.

Match the symbols pictured to the correct electrical device by entering the appropriate letter in each blank space.

_____14. Foot switch

_____15. Silicon controlled rectifier

_____16. NC limit switch

_____17. Pilot light

_____18. Solenoid

_____19. Liquid level switch

_____20. Single-phase motor

_____21. Three-phase motor

_____22. Temperature switch

_____23. Pressure or vacuum switch

_____24. Flow switch

_____25. Control transformer

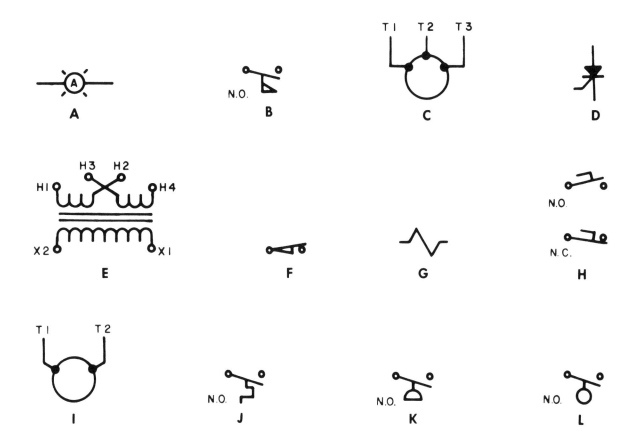

WORKSHEET 3-1

Name _____   Class _____   Date _____

Complete each line diagram according to the circuit information given below. Use standard lettering, numbering, and coding information. Connecting lines should be straight and the circuits neatly drawn.

Circuit 1: Complete the line diagram so that any of three start pushbuttons will start the motor and any one of three stop pushbuttons will stop the motor. This circuit must also include MEMORY so that the motor will remain running after any start pushbutton is pressed and released.

Circuit 2: Redraw Circuit 1, adding two pilot lights. The red pilot light is to be on any time the motor is on, and the green light is to be on any time the motor is off.

Circuit 3: Redraw Circuit 2, adding a selector switch that can be used to place the circuit in a "jog" or "run" position.

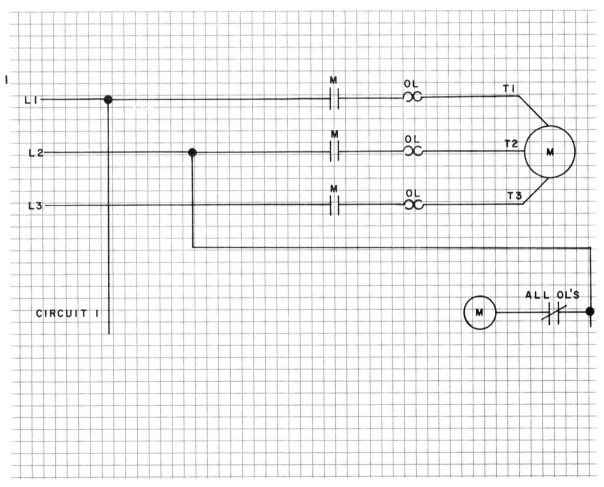

Continued

13

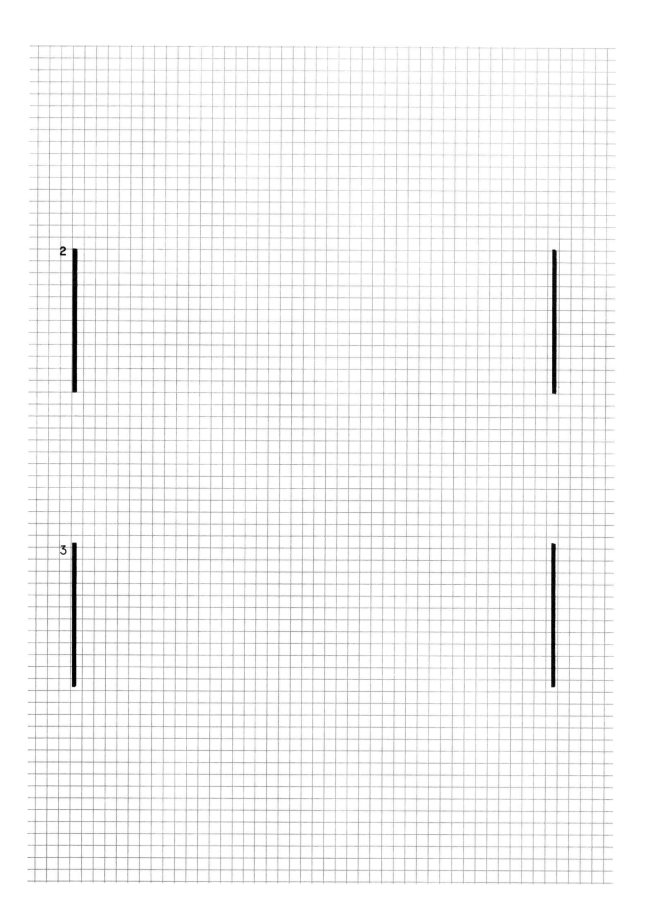

WORKSHEET 3-2

Score _____

Name _____ Class _____ Date _____

Complete the line diagrams according to the circuit information given below. Use standard lettering, numbering, and coding information. Connecting lines should be straight and the circuits neatly drawn.

Circuit 1: Three magnetic motor starters are to be controlled by a common START-STOP pushbutton station. Interconnect the three starters so that if an overload occurs on any one of the starters, all three starters will automatically be disconnected.

Circuit 2: Three magnetic motor starters are to be controlled by three individual START-STOP pushbutton stations. Add to this circuit a master stop that will stop all three starters when pressed. When the master stop is not used, the starters can be individually stopped by each START-STOP station. Each starter must have its own overload protection.

Circuit 3: Redraw Circuit 2, adding a pressure switch that will automatically stop all motors if too high a pressure is reached.

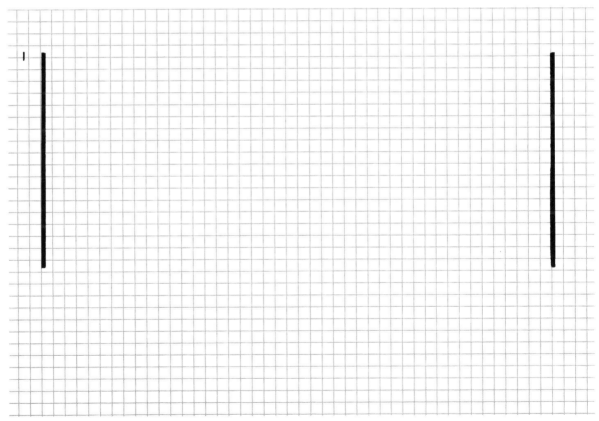

Continued

15

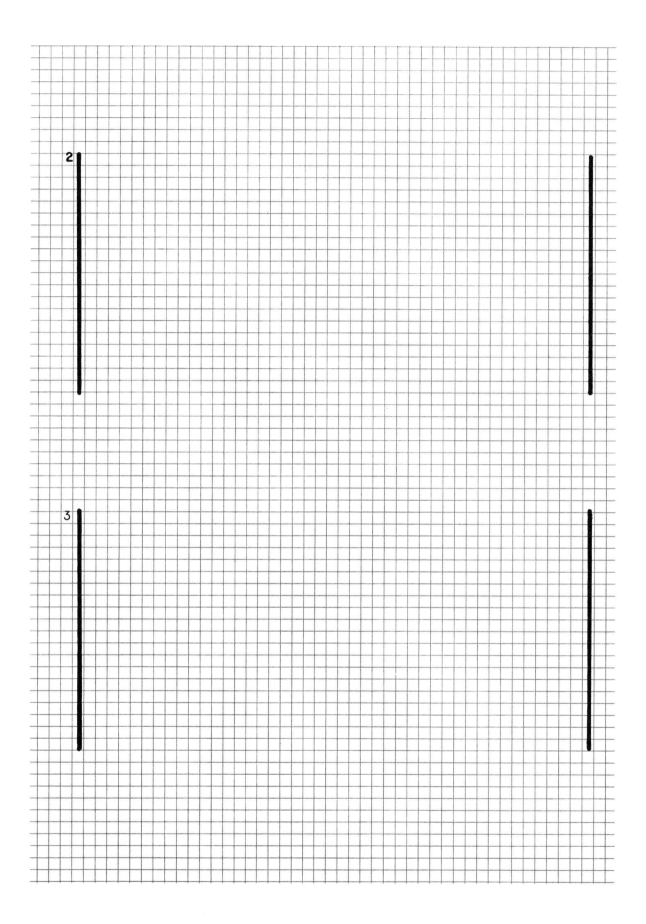

WORKSHEET 3-3

Name _____ Class _____ Date _____

Complete the line diagram on the other side of the page to design a control circuit for a three-belt, three-motored conveyor system that will meet the following operating sequence: Conveyor A feeds bulk material to Conveyor B. Conveyor B feeds the material to Conveyor C. Conveyor C dumps the material. In addition, to prevent pile-ups and assure safe operation, the following conditions must be met.

1. Conveyor A and B will not start unless C is operating.

2. Conveyor A will not start unless B is operating.

3. If Conveyor C stops because of an overload, A and B must stop.

4. If Conveyor B stops because of an overload, A will stop, but C will continue to run.

5. If Conveyor A stops, B and C will continue to run.

6. Only one start and one stop pushbutton must be used to control the conveyor system. However, individual pilot lights will show which conveyors are running.

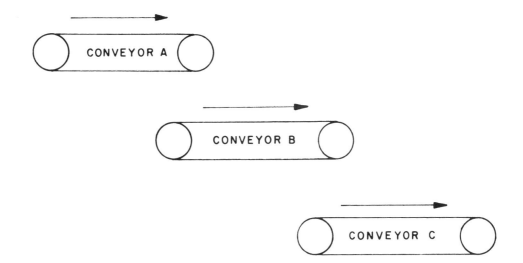

Continued

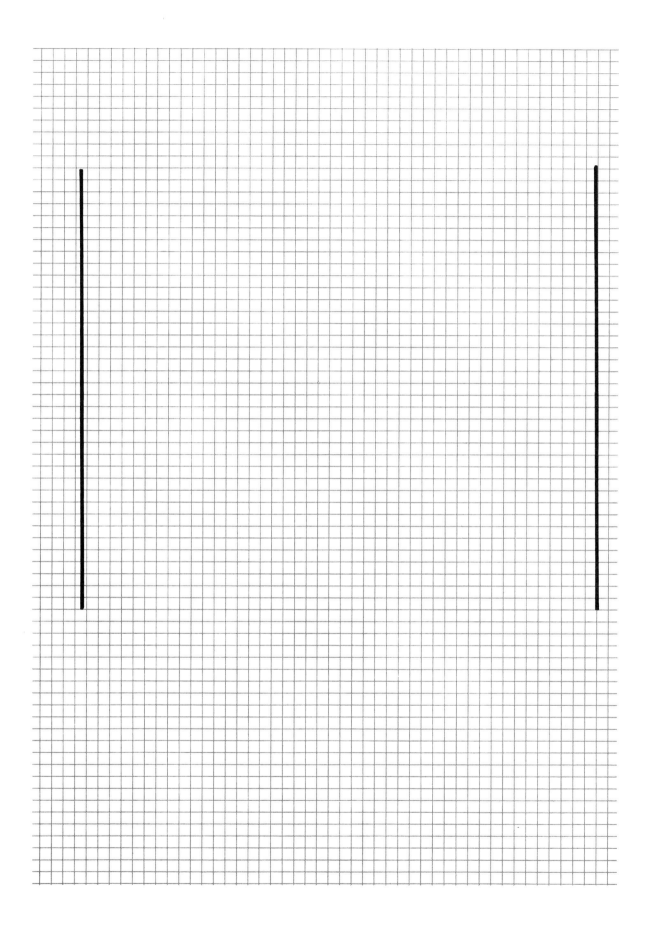

WORKSHEET 3-4

Name _____ Class _____ Date _____

Complete the line diagrams according to the circuit information given below. Use standard lettering, numbering, and coding information. Connecting lines should be straight and the circuits neatly drawn.

Circuit 1: Design a START-STOP-JOG control circuit, using a selector switch to provide the jog-run function. In the jog position, the start buttons should be capable of jogging the motor. In the run position, the circuit should function as a standard START-STOP circuit with memory.

Circuit 2: Illustrate in a circuit how a PUSH-TO-TEST green pilot light can be used to indicate when the starter is energized and to enable the testing of the bulb by simply pushing the color cap on the pushbutton.

Circuit 3: Illustrate how two pushbuttons can be connected for "NAND" logic. The pushbuttons are to control a solenoid.

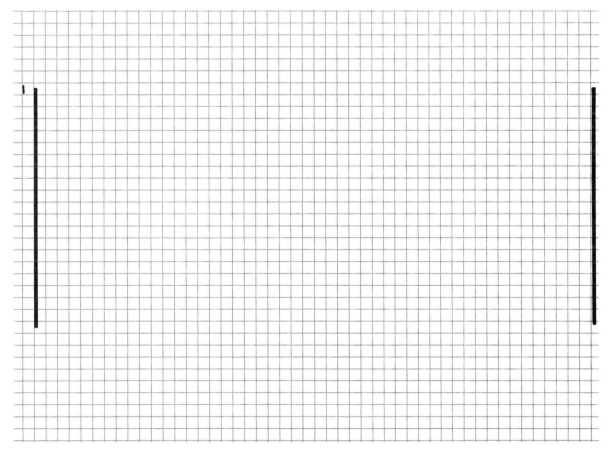

Continued

19

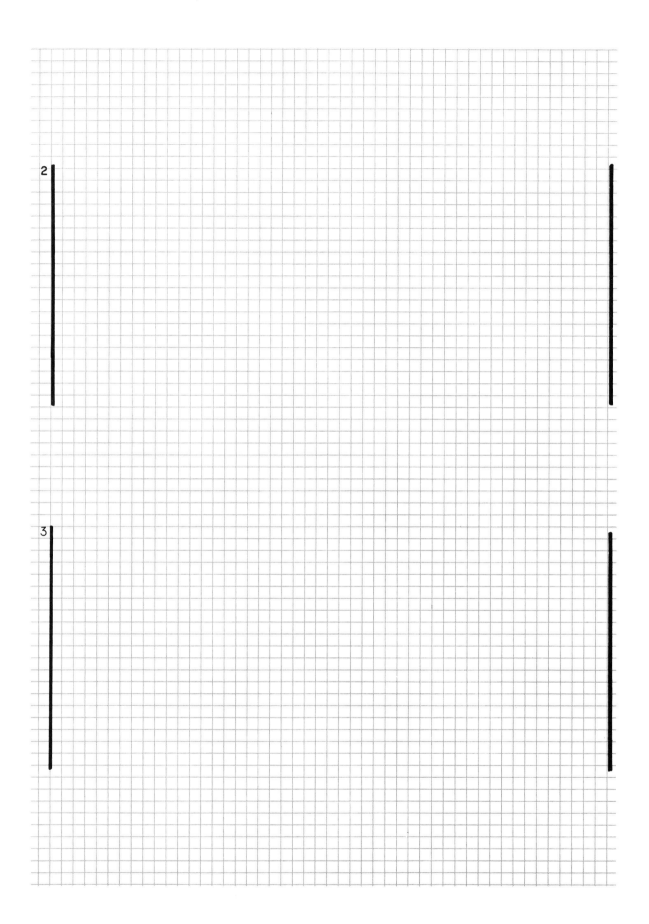

WORKSHEET 3-5

Name _____ Class _____ Date _____

Complete the line diagrams according to the circuit information given below. Use standard lettering, numbering, and coding information. Connecting lines should be straight and the circuits neatly drawn.

Circuit 1: Develop an OR logic circuit according to the conditions stated:
  Signal = one mechanical (limit switch) and one manual (pushbutton).
  Decision = OR logic.
  Action = bell ringing.

Circuit 2: Develop a NOT logic circuit according to the conditions stated:
  Signal = automatic temperature control (temperature switch).
  Decision = NOT logic.
  Action = red pilot light and a heating element activated simultaneously.

Circuit 3: Redraw Circuit 2, adding a second temperature switch so that either switch will activate the loads.

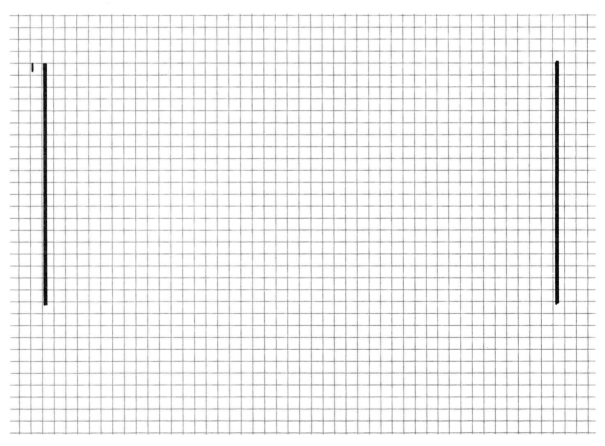

Continued

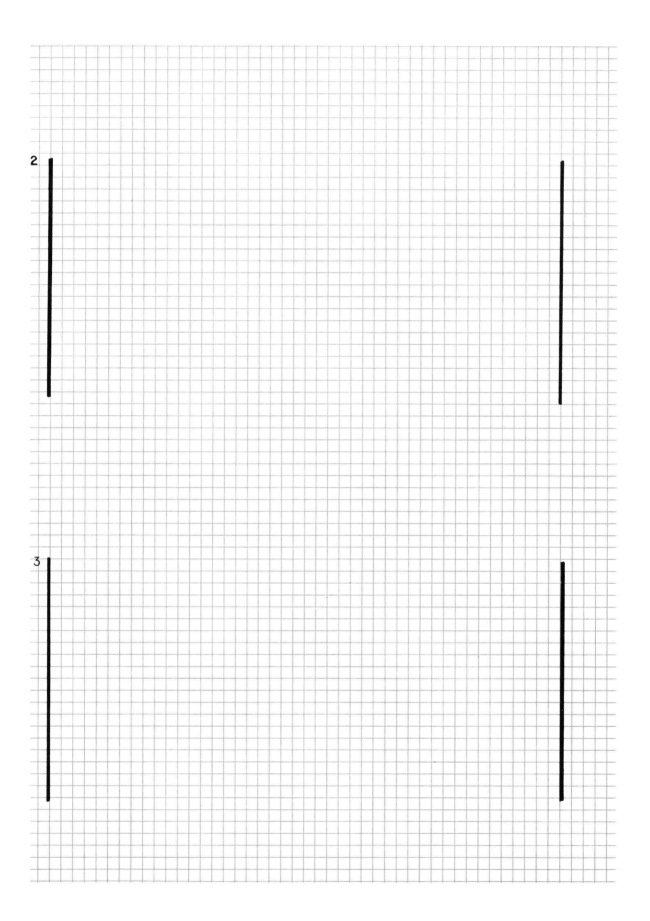

2

3

WORKSHEET 3-6

Name _____ Class _____ Date _____

Complete the line diagrams according to the circuit information given below. Use standard lettering, numbering, and coding information. Connecting lines should be straight and the circuits neatly drawn.

Circuit 1: Develop a circuit with AND logic and with OR logic according to the conditions stated:
Signal = manual (four total: two pushbuttons, two limit switches).
Decision = combination AND logic with OR logic so that at least three devices must be actuated.
Action = siren activated.

Circuit 2: Develop a circuit according to the conditions stated:
Signal = automatic (vacuum switch) and manual (pushbutton).
Decision = MEMORY so that a pushbutton will start the operation and hold until a vacuum switch will stop the operation.
Action = magnetic starter coil.

Circuit 3: Redraw Circuit 2, adding a red pilot light that will indicate when the motor is energized.

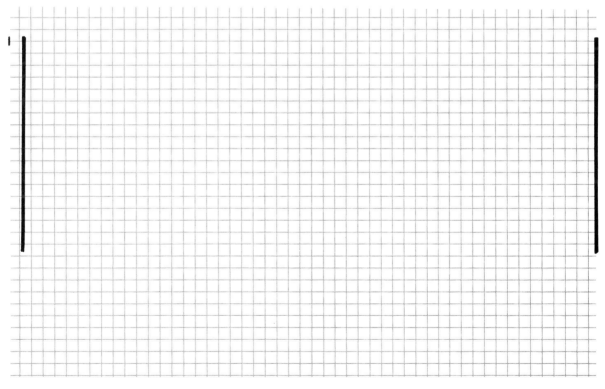

Continued

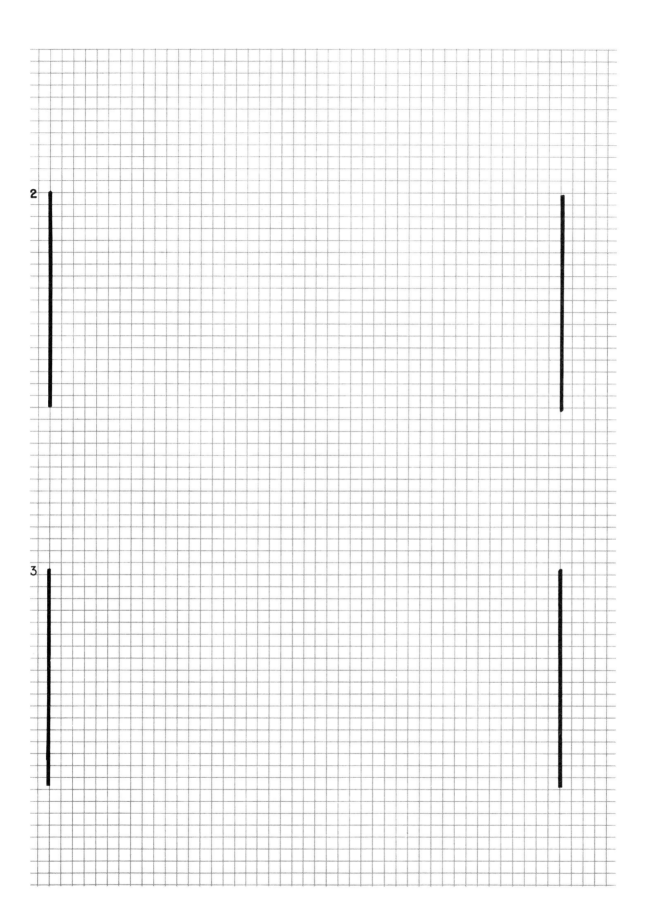

WORKSHEET 3-7

Name _____   Class _____   Date _____

Listed below are ten statements that are illustrated in the ten circuits shown on the other side of this page. Each statement is illustrated by only one of the circuits. Match the statements to the proper circuits by entering the letter of the statement in the blank space to the right of the circuit number.

All the circuits have been drawn with a light (symbolized by "L") to represent the load, whether it is a motor, a bell, a light, or any other load. Also, each switch is illustrated as a pushbutton, whether it is a maintained switch, a momentary switch, a pushbutton, a switch-on target, or any other type of switch.

A. When the regular white light is on in the darkroom, the warning light outside (meaning "do not enter") is off. When the regular white light is off in the darkroom, the warning light outside is on.

B. Since an airplane pilot may panic when in trouble and activate the wrong switch, the switches are connected in such a way that the canopy is ejected first and the pilot second, no matter which switch the pilot activates first.

C. Two guns are connected to individual targets by switches in such a way that when two people compete in firing them, the fastest to fire is always shown.

D. So that no one person can activate the firing of a missile, switches are connected in a way that requires the action of several people.

E. Each person on a panel of three judges has access to a switch that can be activated for a "yes" vote. The switches are connected so that a simple majority vote will be indicated by a light.

F. A security guard monitoring a light panel can tell if the front door, back door, or both front and back doors are open.

G. When the oven is on, an indicating light is also on to warn of danger.

H. The on-off operation of a motor can be controlled from two separate locations.

I. Either one of two machine operators can turn a motor on, but both operators must agree in order to turn the motor off.

J. An old game called "Odds or Evens" is to be played electrically using switches. Three switches are connected in such a way that a light is on every time an odd number of switches is switched, and off every time an even number of switches is switched.

Continued

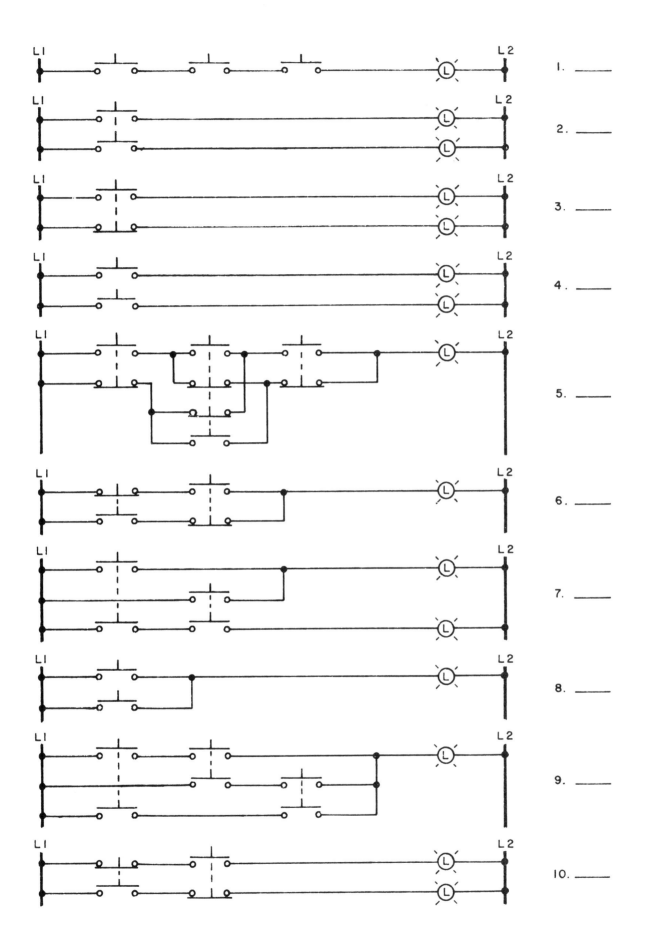

1. _____

2. _____

3. _____

4. _____

5. _____

6. _____

7. _____

8. _____

9. _____

10. _____

TECH-CHEK ✔ 3

Name _____

Class _____ Date _____

Score _____

Introduction to Logic as Applied to Line Diagrams and Basic Control Circuits

Fill in the blanks to complete each statement.

1. When more than one electrical load is connected in a line diagram, the loads are connected in _____ (parallel or series).

2. Control relays, solenoids, and pilot lights are examples of loads connected directly to line _____.

3. A _____ is an example of a load that is connected indirectly to line 2 (L2).

4-5. In a line diagram, pushbuttons, limit switches, and pressure switches are connected between (4)_____ and the (5)_____.

6-8. The control circuit is composed of three basic sections called the (6)_____, the (7)_____, and the (8)_____.

9. The _____ is that section of the control circuit that starts or stops the flow of current by closing or opening the control devices' contacts.

10. The _____ is that section of the circuit that determines what work is to be done and in what order the work is to occur.

11. The _____ is that section of the circuit that contains the load.

12. A _____ (manual, automatic, standard, or mechanical) condition refers to any input into the circuit by a person.

13. A _____ (manual, automatic, standard, or mechanical) condition refers to any input into the circuit by some moving part.

14. A _____ (manual, automatic, standard, or mechanical) condition is one that will respond to changes in a system.

15. The circuit below is an example of _____ control logic.

Continued

16. The circuit below is an example of _____ control logic.

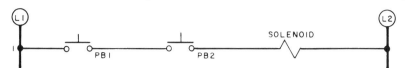

17. The circuit below is an example of _____ control logic.

18. The circuit below is an example of _____ control logic.

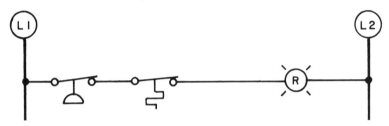

19. The circuit below is an example of _____ control logic.

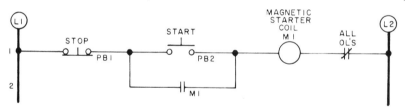

20. The circuit below is an example of a combination of _____ and

_____ and _____ logic.

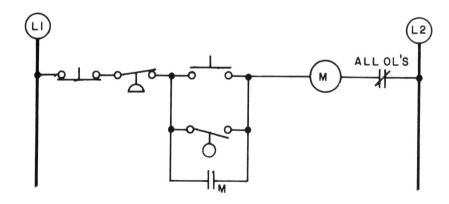

WORKSHEET 4-1

Name _____ Class _____ Date _____

Below each illustration write the type of NEMA enclosure that would be best for the location shown.

1. _____

2. _____   Continued

3. _____

4. _____

WORKSHEET 4-2

Name _____   Class _____   Date _____

Referring to Data Sheet A (see back of workbook), complete the ten basic switching arrangements that are possible with a dual element coil and dual voltage power supply. Draw a schematic diagram of each circuit to the right of the switching arrangement. Your lines should be straight and the circuits neatly drawn.

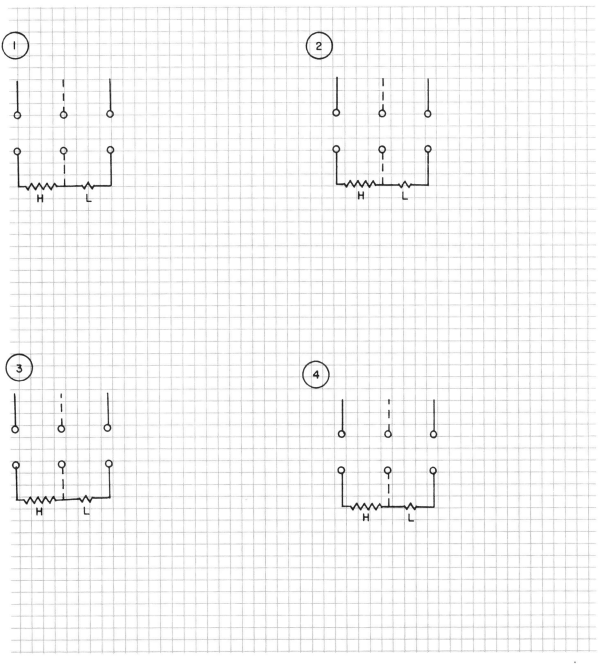

Continued

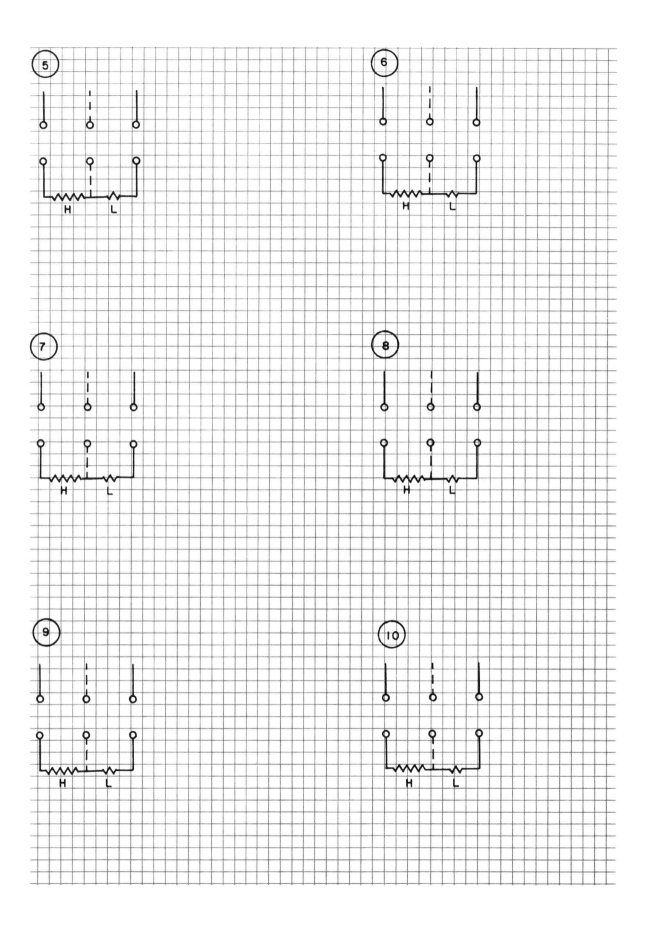

TECH-CHEK ✔ 4

Name _____

Class _____ Date _____

Score _____

AC Manual Contactors and Motor Starters

Select the best way to complete each statement and circle a, b, c, or d. Do not circle more than one choice for each statement.

1. When electric motors were first introduced, the starting and stopping of the motor was done through a
a. magnetic motor starter.
b. magnetic contactor.
c. knife switch.
d. solenoid switch.

2. One of the advantages of using a double break contact instead of a single break contact is that the double break contact
a. has a higher contact rating in a smaller space.
b. has a higher contact rating when enclosed in a steel case.
c. is not made of soft copper.
d. does not present an electrical hazard.

3. In manual contactors, external electrical connections are made indirectly to the fixed contacts through
a. the knife blades.
b. the movable contacts.
c. saddle clamps.
d. the springs.

4. Arc hoods not only insulate each set of contacts from each other but also
a. help move the movable contacts.
b. help make a better electrical connection.
c. help close the contacts faster.
d. help contain and quench the arcs.

Fill in the blanks to complete each statement.

5. Because of the disadvantages of using copper for switching, copper contacts have been

replaced with _____ alloy contacts.

6. Unlike the line diagram, the _____ diagram is intended to show as

closely as possible the actual connection and placement of all component parts.

7. When two contactors are connected in such a way that both sets of contacts cannot be

closed at the same time, the contactors are _____ interlocked.

8. The primary difference between a contactor and a starter is the addition of

_____ protection to the starter.

Continued

9. When a motor is loaded so heavily that the motor shaft cannot turn, a condition called

_____ exists.

10-12. The three stages that a motor must go through in normal operation are:

(10)_____ (11)_____ and

(12)_____

13-14. To protect a motor against very high currents, short circuits or a ground, a

(13)_____ or (14)_____ is used.

15. A time delay device called the _____ is used to protect a motor

while it is running.

16-17. A 120-volt, single-phase power source has (16)_____ hot wire(s), and

(17)_____ neutral wire(s).

18-19. A 230-volt, single-phase power source has (18)_____ hot wire(s), and

(19)_____ neutral wire(s).

20-21. A three-phase power source has (20)_____ hot wire(s) and

(21)_____ neutral wire(s).

22. A NEMA type _____ enclosure is intended for outdoor as well as

indoor use, since it provides some protection against windblown dust and rain, splashing water,

and hose-directed water.

WORKSHEET 5-1

Name _____ Class _____ Date _____

Complete the line diagram on the other side of the page by referring to the illustration below, which shows a double solenoid valve controlling the cylinder.

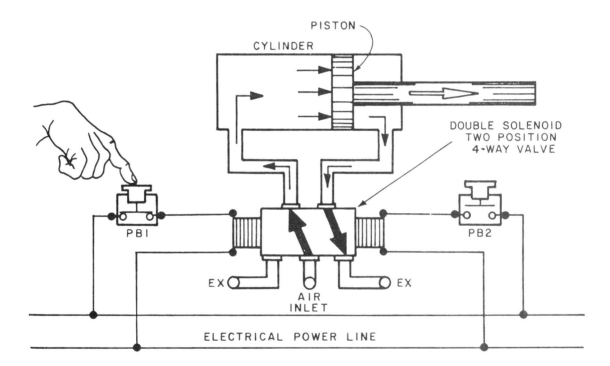

Continued

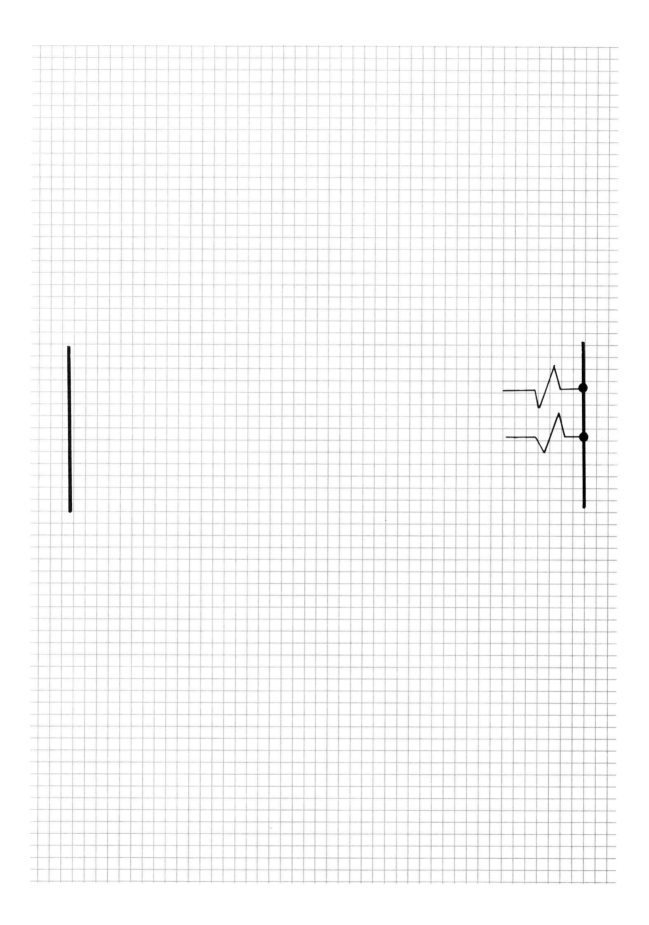

WORKSHEET 5-2

Score _____

Name _____ Class _____ Date _____

Complete the line diagrams according to the circuit information given below. Use standard lettering, numbering, and coding information. Connecting lines should be straight and the circuits neatly drawn.

Circuit 1: In Worksheet 5-1, the pushbutton must be held down until the cylinder is fully advanced. Change this circuit so that when the start pushbutton is pressed and released, the cylinder will advance. Also include in this circuit a limit switch that will automatically stop the cylinder when the cylinder is fully advanced. To aid you in accomplishing your goal, a contactor is provided in parallel with the solenoid. This contactor is capable of controlling NO and NC contacts.

Circuit 2: Redraw Circuit 2, adding a second pushbutton to return the cylinder and a second limit switch to stop the cylinder when fully retracted. This circuit should be drawn to illustrate the circuit in the retracted position at rest. Mark the pushbuttons "forward" and "return."

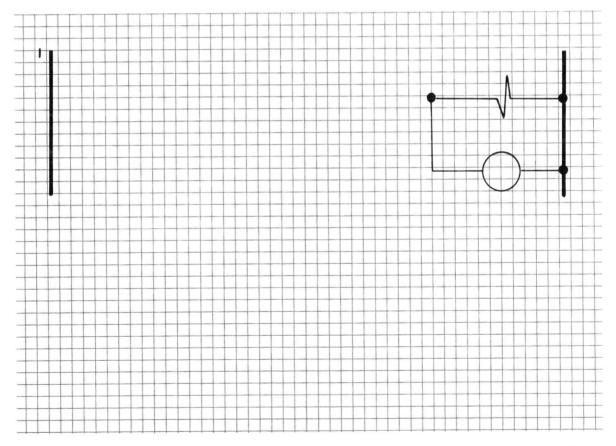

Continued

37

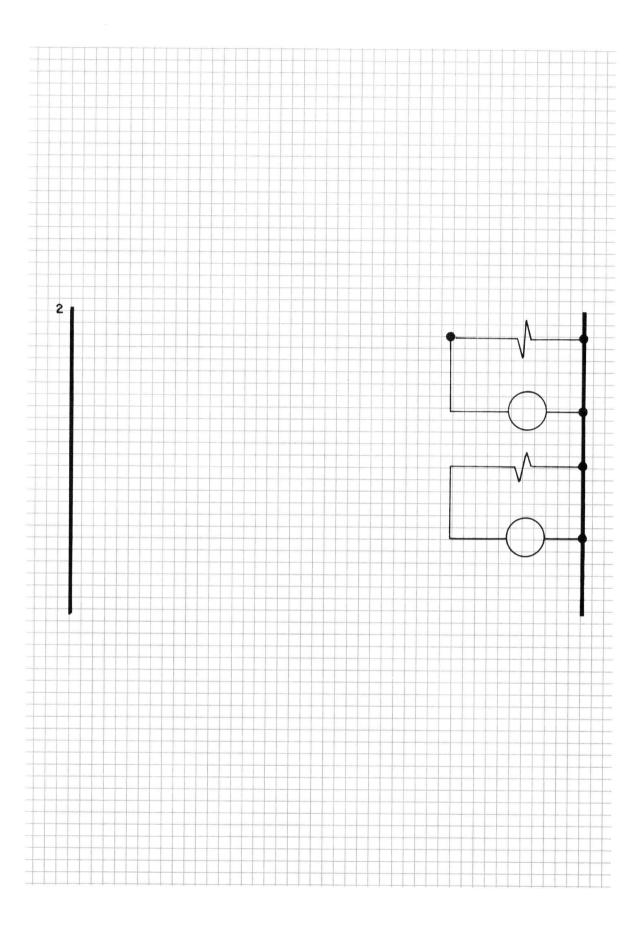

WORKSHEET 5-3

Name _____ Class _____ Date _____

Referring to Data Sheet C, note that limit switch 1 (LS1) is normally open (held closed) because the cylinder is retracted. Assume that no MEMORY is required for either solenoid. Complete the line diagram showing how a selector switch could be used to determine manual or automatic control of a cylinder. When the selector switch is in the "automatic" position, the cylinder should continuously cycle back and forth automatically. When the selector switch is in the "manual" position, the cylinder should advance only when a pushbutton is pressed and held, and retract only when a second pushbutton is pressed and held. Use standard lettering, numbering, and coding information. Connecting lines should be straight and the circuits neatly drawn.

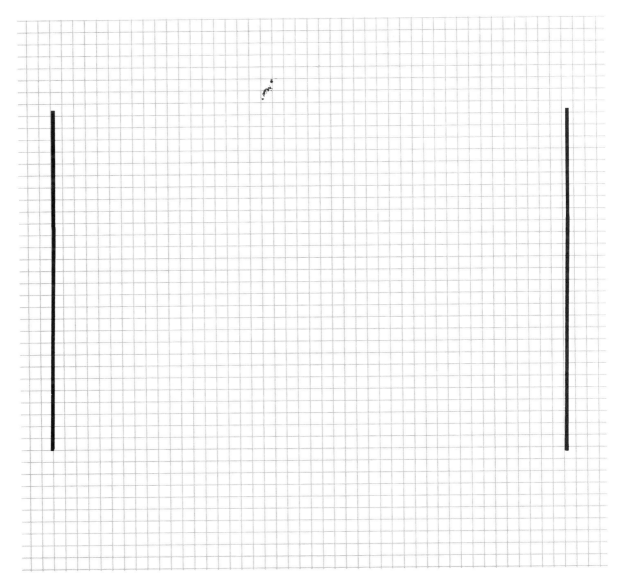

Continued

WORKSHEET 5-4

Name _____ Class _____ Date _____

Redraw the circuit of Worksheet 5-3, adding an emergency stop pushbutton that will stop the solenoids from being energized until a reset pushbutton is activated.

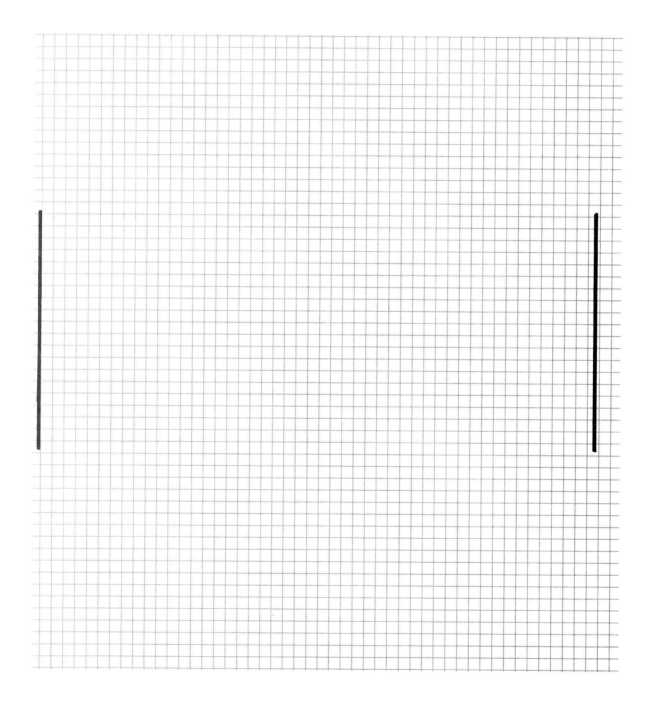

TECH-CHEK **5**

Name _____

Class _____ Date _____

Score _____

Magnetism and Magnetic Solenoids

Select the best way to complete each statement and circle a, b, c, or d. Do not circle more than one choice for each statement.

1. The strength of an electromagnet can be increased by
   a. increasing the voltage.
   b. increasing the number of turns of wire.
   c. inserting an iron core.
   d. all of the above.

2. The solenoid armature is made from thin laminated pieces to help reduce
   a. air caps.
   b. eddy currents.
   c. slow action.
   d. cost.

3. The reason for a small air gap left in the iron core armature circuit is to
   a. prevent eddy currents.
   b. prevent the armature from staying in a sealed position.
   c. reduce cost.
   d. prevent heat build-up in the armature circuit.

4. The reason for adding a shading coil to the armature core is to
   a. reduce air gaps.
   b. keep the armature firmly seated.
   c. prevent eddy currents.
   d. prevent the armature from staying in a sealed position.

5. Excessively noisy solenoids may be a result of
   a. broken shading coil.
   b. voltage too low.
   c. dirt, rust, or fillings on the magnetic face.
   d. all of the above.

Fill in the blanks to complete each statement.

6. _____ magnets are magnets which can retain their magnetism after a

magnetizing force has been removed.

7. _____ magnets are magnets which have extreme difficulty in retaining

any magnetism after the magnetizing force has been removed.

8. The current that a solenoid coil draws when first turned on is called the

_____.

9. The current that a solenoid coil draws after the armature circuit is closed is called the

_____.

Continued

10. The amount of force a solenoid can deliver is usually rated in _____.

11. The number of times a solenoid can operate in a given time (usually per minute) is called

the _____.

Identify the different types of solenoids pictured by entering the appropriate letter in each blank space.

_____12. Vertical action type

_____13. Horizontal action type

_____14. Clapper type

_____15. Bellcrank type

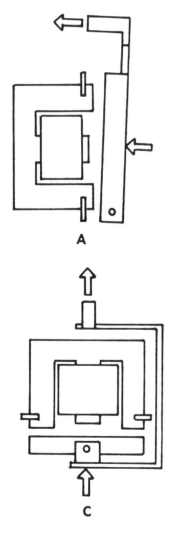

A

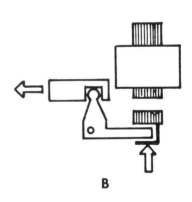

B

C

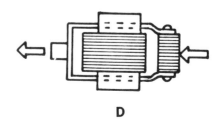

D

WORKSHEET 6-1

Score _____

Name _____ Class _____ Date _____

Referring to Data Sheet B (see back of workbook), assign a number to each wire in the circuit shown below. Start in the upper lefthand corner and move to the right, line by line. Write each number directly above the wire.

This circuit is designed in such a way that two magnetic starters are operated by two START-STOP stations with a common emergency stop. Shown below the circuit are two pilot lights, labeled R for red and G for green. After you have numbered every wire in the circuit, assign the proper numbers to the wires of the two pilot lights so that when they are connected into the circuit, the red pilot light will glow when magnetic starter M-1 is on and the green will glow when magnetic starter M-2 is on.

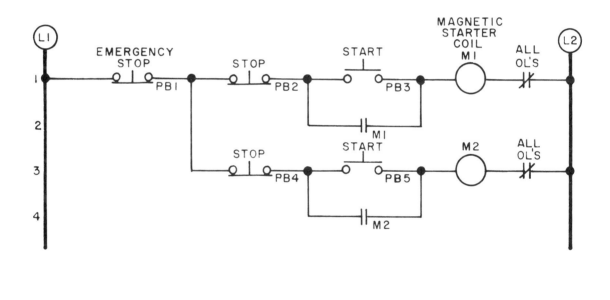

WORKSHEET 6-2

Name _____ Class _____ Date _____

Referring to Data Sheet C (see back of workbook), assign a number to each wire in the circuit shown below. Start in the upper lefthand corner and move to the right, line by line. Write each number directly above the wire.

This circuit is designed with pushbuttons arranged for a sequence control of two starters. Shown below the circuit are a footswitch and a pushbutton. After you have numbered every wire in the circuit, assign the proper numbers to the wires of the footswitch so that when it is connected into the circuit, it will also start coil M1. Assign the proper numbers to the wires of the pushbutton so that when it is connected into the circuit, it will also start coil M2.

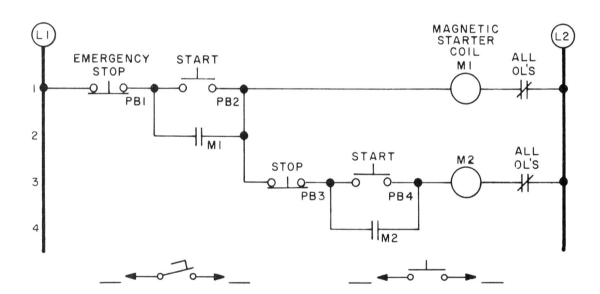

WORKSHEET 6-3

Name _____ Class _____ Date _____

Complete the wiring diagram according to the line diagram shown below of a START-STOP station with MEMORY. Connecting lines should be straight and the circuit neatly drawn. Do not make any wire splices or additional terminal connections on the wiring diagram. All connections must run from terminal screw to terminal screw.

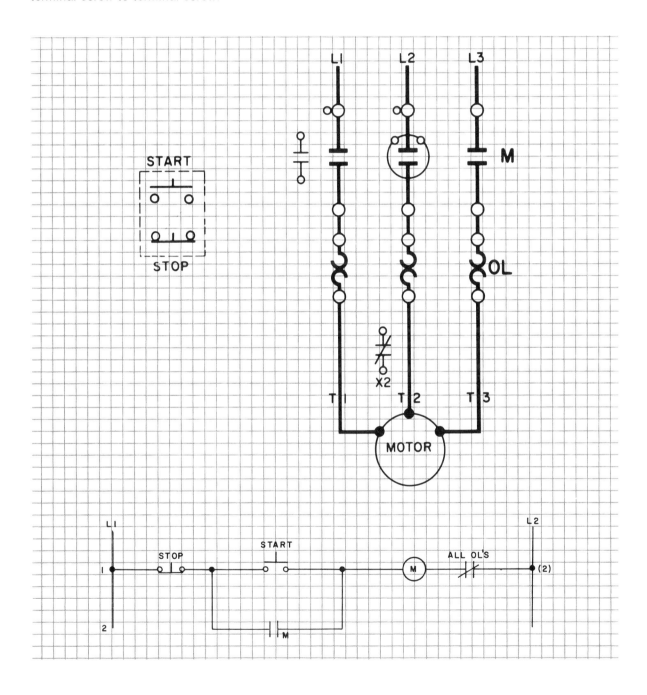

WORKSHEET 6-4

Score _____

Name _____ Class _____ Date _____

Complete the wiring diagram according to the line diagram shown below of a START-STOP station with MEMORY and with a pilot light that turns on when the motor is not running. There is overload protection for the motor.

Your connecting lines should be straight and the circuit neatly drawn. Do not make any wire splices or additional terminal connections on the wiring diagram. All connections must run from terminal screw to terminal screw.

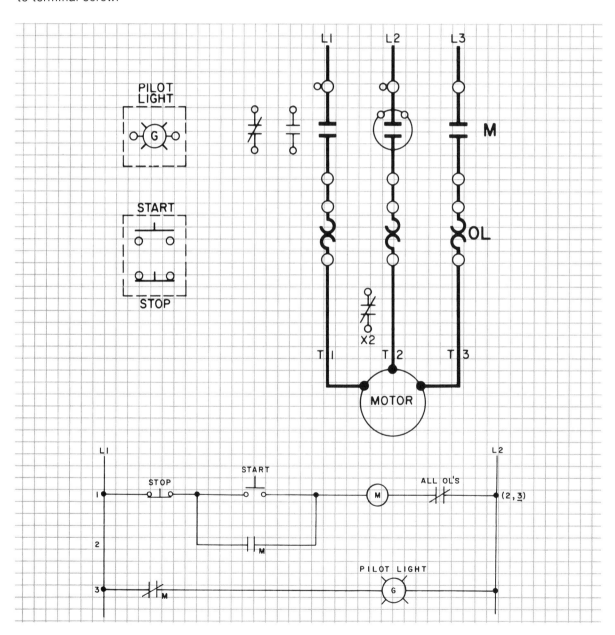

WORKSHEET 6-5

Score _____

Name _____ Class _____ Date _____

Complete the wiring diagram according to the line diagram shown below of three START-STOP stations with MEMORY, controlling a single starter. There is overload protection for the motor.

Your connecting lines should be straight and the circuit neatly drawn. Do not make any wire splices or additional terminal connections on the wiring diagram. All connections must run from terminal screw to terminal screw.

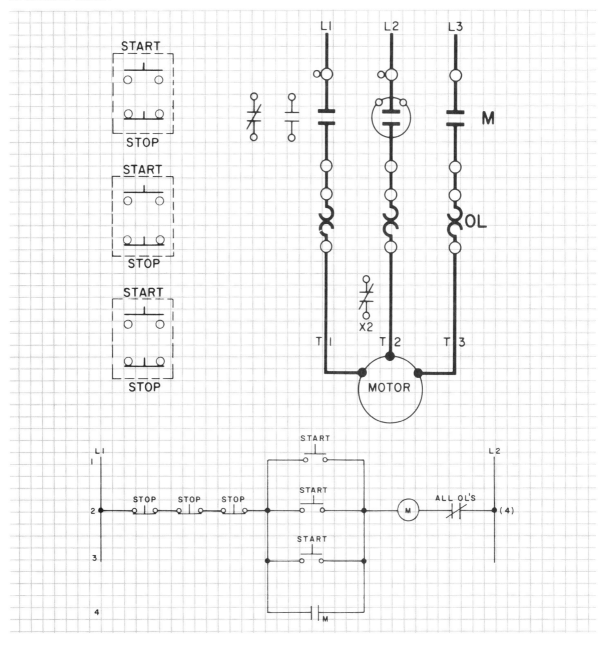

WORKSHEET 6-6

Score _____

Name _____ Class _____ Date _____

Complete the wiring diagram according to the line diagram shown below of a START-STOP station with MEMORY, controlling two starters. The starters are wired so that if a maintained overload occurs on either one of the starters, both will be automatically disconnected from the line. This is accomplished by wiring the holding circuit of each starter through the auxiliary contacts of the other. The circuit also provides for a sequence start of each starter and motor in order to avoid the problems that could arise from both motors starting at once.

Your connecting lines should be straight and the circuit neatly drawn. Do not make any wire splices or additional terminal connections on the wiring diagram. All connections must run from terminal screw to terminal screw.

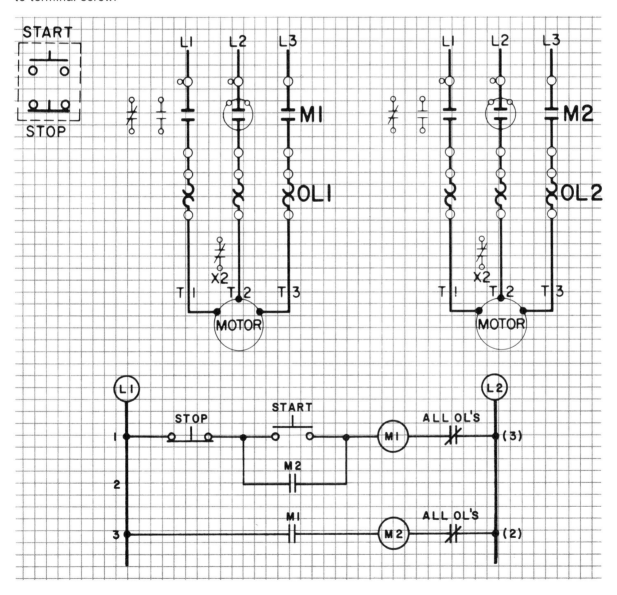

WORKSHEET 6-7

Score _____

Name _____ Class _____ Date _____

Complete the wiring diagram on the other side of the page according to the line diagram shown below of two separate START-STOP stations with MEMORY, controlling two separate starters. A master stop pushbutton is included to turn off both starters. The overload relays on both starters are wired in series so that when a maintained overload occurs in one, both will drop out.

Your connecting lines should be straight and the circuit neatly drawn. Do not make any wire splices or additional terminal connections on the wiring diagram. All connections must run from terminal screw to terminal screw.

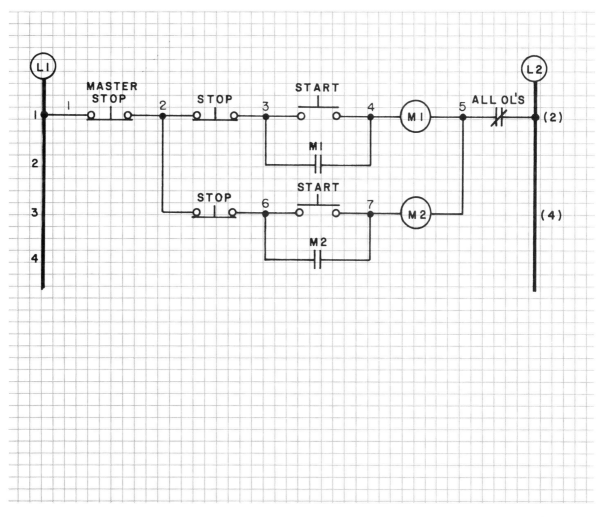

Continued

49

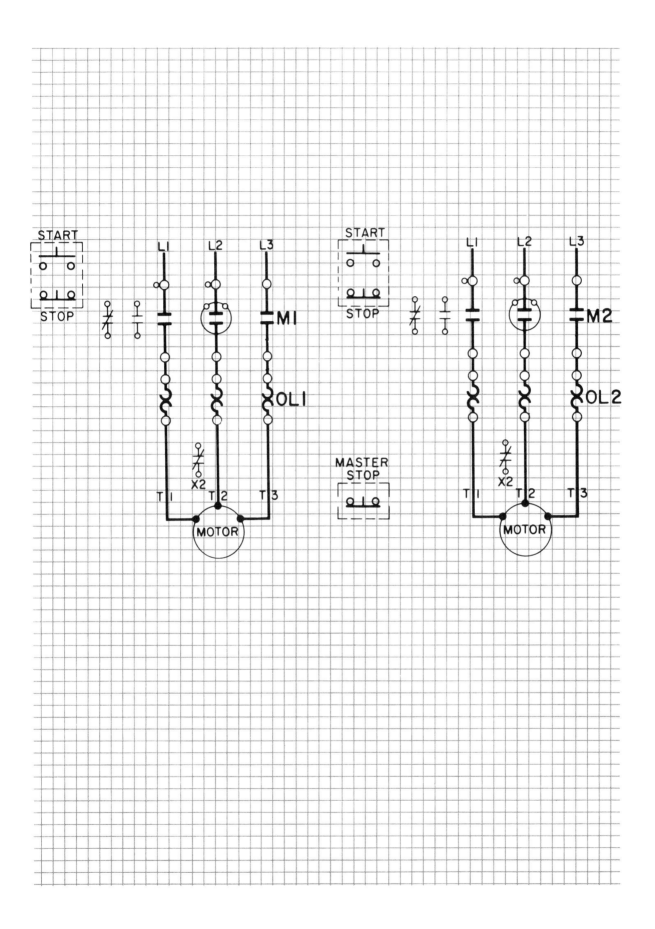

START
STOP

L1 L2 L3

M1

OL1

X2

T1 T2 T3

MOTOR

START
STOP

MASTER
STOP

L1 L2 L3

M2

OL2

X2

T1 T2 T3

MOTOR

WORKSHEET 6-8

Score _____

Name _____ Class _____ Date _____

Referring to Data Sheet C, indicate how the equipment illustrated there would be wired if confined to the conduit connection and enclosure arrangement shown below. Power feed should be through the START-STOP enclosure. All wire splices must be inside the enclosure and not within the conduit. Do not make any wire splices that are not necessary.

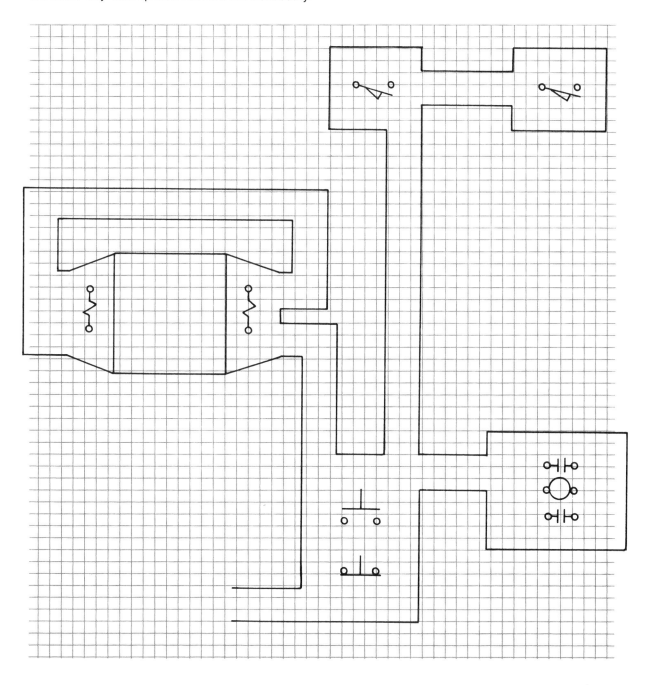

51

WORKSHEET 6-9

Score _____

Name _____ Class _____ Date _____

Referring to Data Sheet C, indicate how the equipment illustrated there would be wired if confined to the conduit connection and enclosure arrangement shown below. Power feed should be through the limit switch enclosure. All wire splices must be inside the enclosure and not within the conduit. Do not make any wire splices that are not necessary.

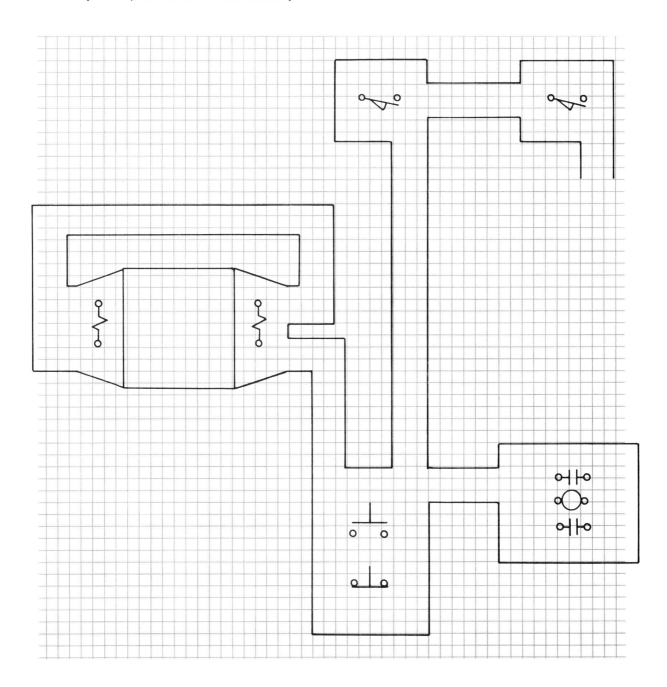

WORKSHEET 6-10

Name _____ Class _____ Date _____

Referring to Data Sheet C, indicate how the equipment illustrated there would be wired if confined to the conduit connection and enclosure arrangement shown below. Power feed should be through the control relay enclosure. All wire splices must be inside the enclosure and not within the conduit. Do not make any wire splices that are not necessary.

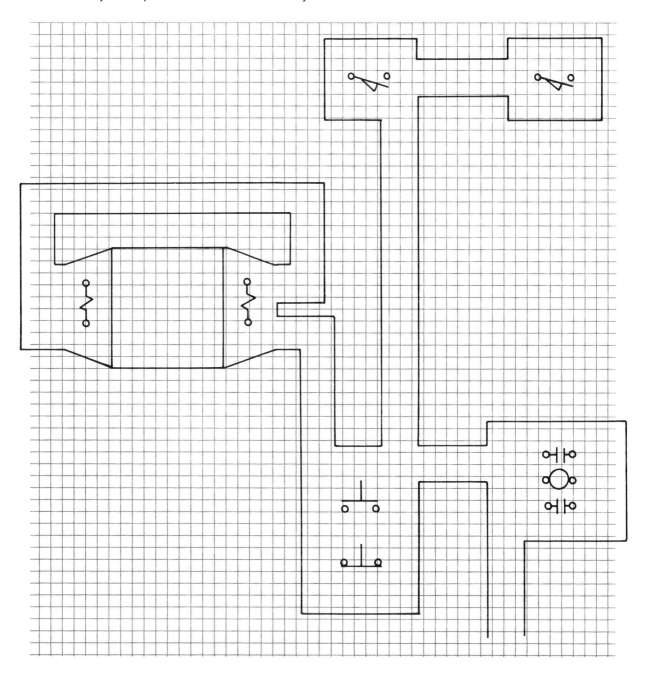

WORKSHEET 6-11

Name _____ Class _____ Date _____

Referring to Data Sheet C, indicate how the equipment illustrated there would be wired if confined to the conduit connection and enclosure arrangement below. Power feed should be through the solenoid valve. All wire splices must be inside the enclosure and not within the conduit. Do not make any wire splices that are not necessary.

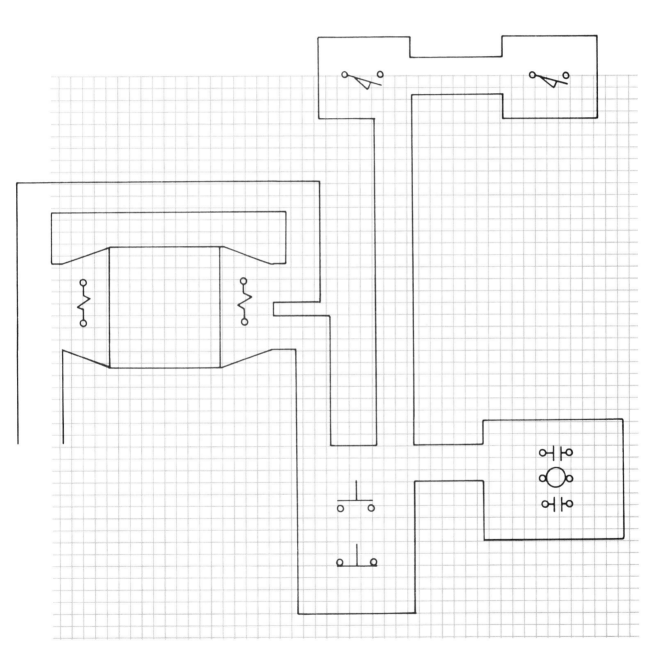

Select the best way to complete each statement and circle a, b, c, or d.

1. When a maintained contact switch such as a float switch is used to control a contactor or starter, and is connected in such a way as to disconnect the load when the power fails and reconnect the load when power is restored, the control circuit is referred to as a
    a. two-wire control.
    b. three-wire control.

2. When a momentary contact switch such as pushbutton is used to control a contactor or starter, and is connected in such a way as to disconnect the load when the power fails and not reconnect the load when power is restored, the control circuit is referred to as a
    a. two-wire control.
    b. three-wire control.

3. When using a contactor to control a DC load, you are required to break
    a. one power line.
    b. two power lines.

4. When using a contactor to control a three-phase load, you are required to break
    a. two power lines.
    b. three power lines.

5. The most difficult arcs to extinguish on a set of contacts are
    a. AC arcs.
    b. DC arcs.

6. A way to help confine, divide, and extinguish the arc for each set of contacts is to use
    a. arc chutes.
    b. overloads.
    c. a current transformer.
    d. a transient suppression module.

7. A way to provide a magnetic field that will help move the contacts some distance from each other as quickly as possible is to use
    a. silver on the contacts.
    b. a current transformer.
    c. blow-out coils.
    d. overloads.

8. As the NEMA number of a contactor or starter increases (size 1, 2, etc.), the power rating of the contactor or starter
    a. increases.
    b. decreases.

9. The current rating of a contactor or starter is the rating for
    a. each individual contact in the contactor.
    b. the whole contactor divided by the number of contacts.

Continued

Fill in the blanks to complete each statement.

10. The main difference between a contactor and a motor starter is the addition of the

_____ to the motor starter.

11-12. The overload protection of motors is accomplished by the overload relay. Two of the different

kinds of overload relays used to protect motors are (11)_____

and (12)_____.

13-15. In selecting the overloads for a starter motor combination, the three major things to know are

the (13)_____, (14)_____ and

(15)_____.

16. If a motor has a service factor of 1.25 and a current rating of 10 amps, the amount of current

it can safely draw for a short period of time is _____ amps.

17. On a three-phase motor installation, the number of individual overloads required

is _____.

18. An inherent motor protector is designed to protect the motor from _____.

19-20. Two optional modifications that can be added to a contactor or motor starter are

(19)_____ and (20)_____.

WORKSHEET 7-1

Score _____

Name _____ Class _____ Date _____

Number each line diagram below, and establish a cross-reference for each control device that uses NO and NC contacts in the line diagram.

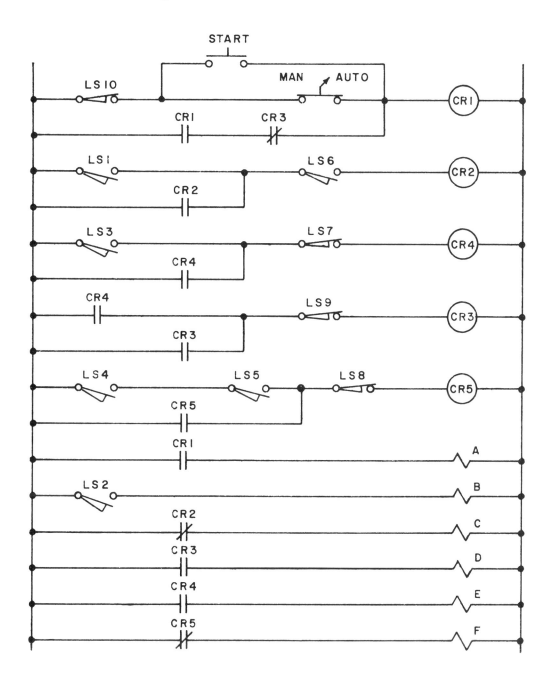

WORKSHEET 7-2

Name _____ Class _____ Date _____

Number each line diagram below, and establish a cross-reference for each control device that uses NO and NC contacts in the line diagram. This circuit provides for a two-hands "no tie down operation". The operator has two seconds to press each pushbutton. If any pushbutton is tied down or held in position, the circuit will not operate. This circuit is used as a protection device on equipment like shears and punch presses.

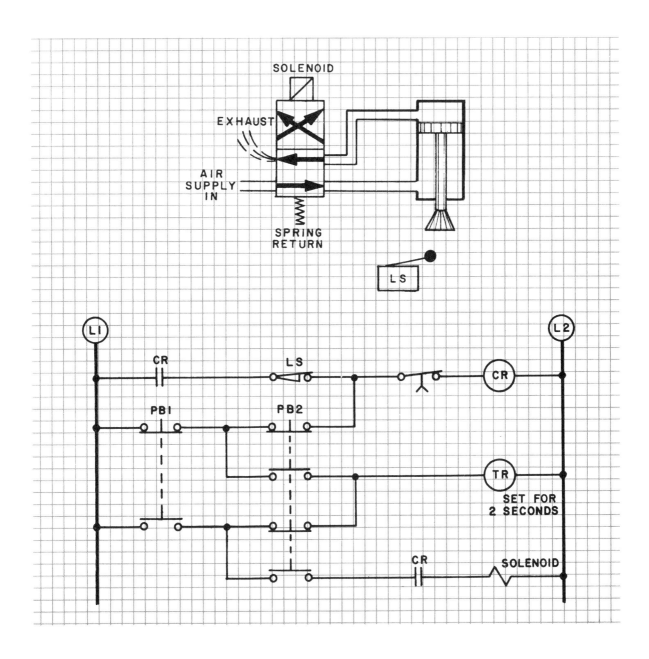

WORKSHEET 7-3

Name _____ Class _____ Date _____

Complete the wiring diagram according to the line diagram shown below of a START-STOP station with MEMORY, controlling two motors. A time-delay relay is provided to prevent both motors from starting at the same time. An overload in Motor 1 will shut down the entire circuit; however, an overload in Motor 2 will affect only Motor 2. This circuit is often used because incoming power line limitations in some areas prohibit the starting of two or more motors at the same time.

Your connecting lines should be straight and the circuit neatly drawn. Do not make any wire splices or additional terminal connections on the wiring diagram. All connections must run from terminal screw to terminal screw.

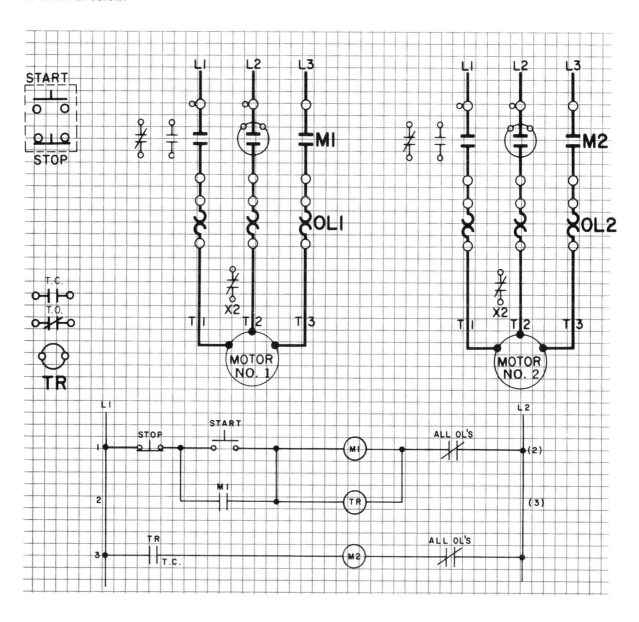

WORKSHEET 7-4

Name _____ Class _____ Date _____

Complete the wiring diagram according to the line diagram shown below, in which Motor 2 starts and runs for a short time after Motor 1 has stopped. An overload in Motor 1 will shut down the entire circuit; however, an overload in Motor 2 will affect only Motor 2. This circuit is used when a second motor is to run for a short time after the controlling motor has stopped, as in a cooling fan or pump.

Your connecting lines should be straight and the circuit neatly drawn. Do not make any wire splices or additional terminal connections on the wiring diagram. All connections must run from terminal screw to terminal screw.

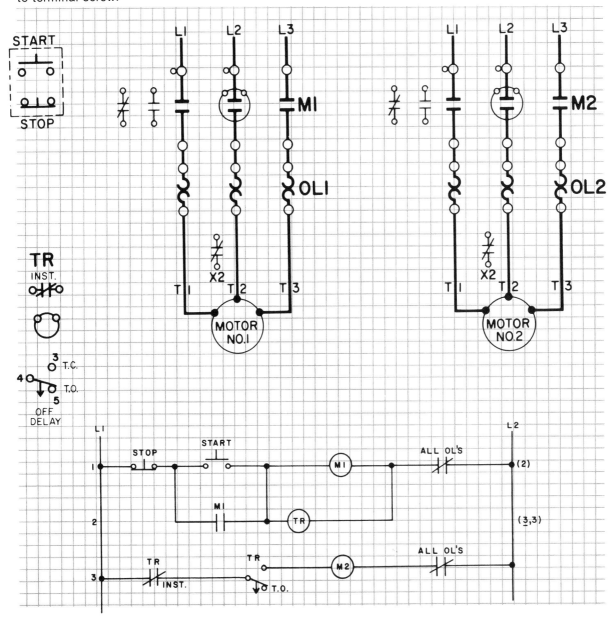

WORKSHEET 7-5

Score _____

Name _____ Class _____ Date _____

Referring to Data Sheet D, complete the wiring diagram for the six circuits below based on their established coding system for the load.

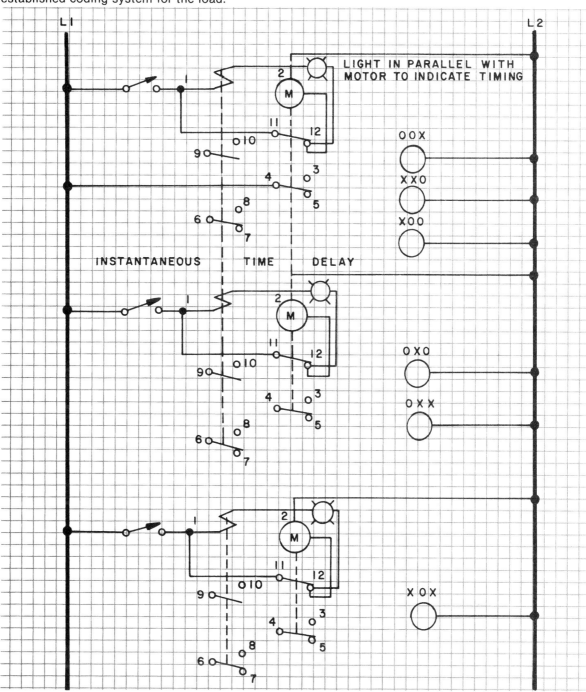

WORKSHEET 7-6

Name _____ Class _____ Date _____

Referring to Data Sheet E, complete the timing diagram below. A conveyor system is to be installed in which the first conveyor (controlled by M1) is turned on by a standard start pushbutton with MEMORY. After the first conveyor has run for one minute, a second conveyor (controlled by M2) turns on automatically. Both conveyors run until a standard stop pushbutton causes both to stop. Each conveyor motor should have independent overload protection.

Use standard lettering, numbering, and coding information. Connecting lines should be straight and the circuits neatly drawn.

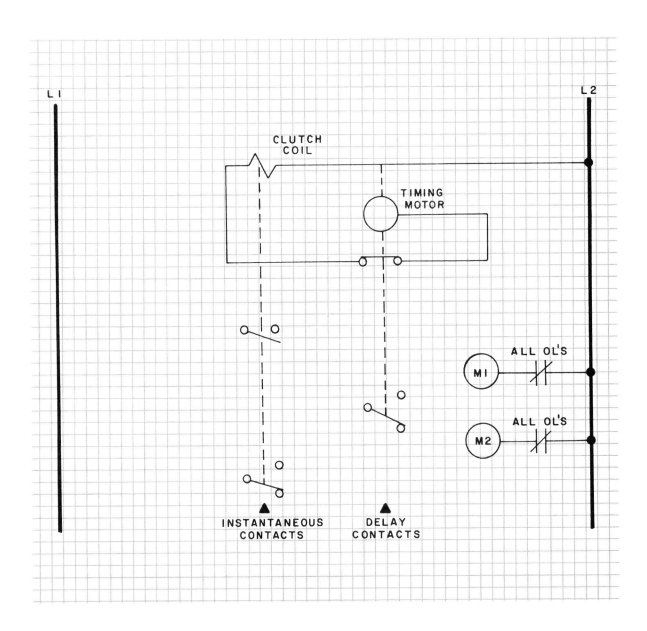

WORKSHEET 7-7

Name _____ Class _____ Date _____

Referring to Data Sheet E, complete the timing diagram below. A time control is to be installed to operate a sandblasting machine. When the operator places the part to be sandblasted in the machine, closes the door, and turns on a toggle switch, the part is to be automatically sandblasted for 30 seconds. The sandblaster is powered by a motor (controlled by M1). During the sandblasting a red light should come on, indicating danger. At all other times a green light should be on, indicating it is safe to open the door. Overload protection should be provided for the motor.

Use standard lettering, numbering, and coding information. Connecting lines should be straight and the circuits neatly drawn.

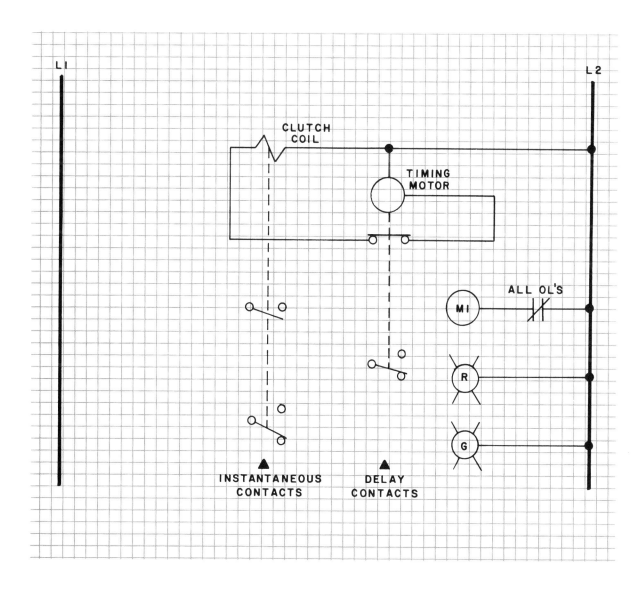

WORKSHEET 7-8

Name _____ Class _____ Date _____

Referring to Data Sheet E, complete the timing diagram. Redraw the sandblasting circuit on Worksheet 7-7 in such a way that a pushbutton could be built into the door of the machine to turn off the sandblasting operation any time the door is open.

Use standard lettering, numbering, and coding information. Connecting lines should be straight and the circuit neatly drawn.

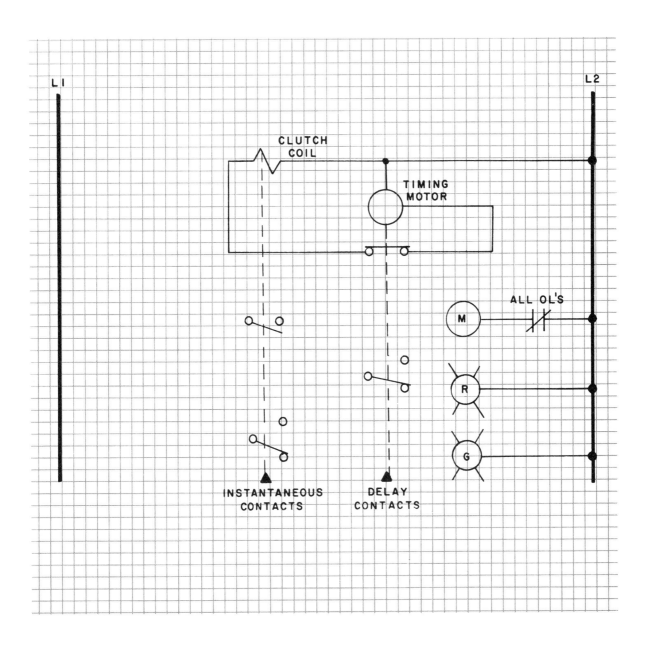

**Time Delay and Logic Applied to More Complex Line
Diagrams and Control Circuits**

Fill in the blanks to complete the statements.

1-3.   The three basic types of timers used in control circuits are the (1)_____,

(2)_____ and (3)_____.

4-6.   The code that is added to a load controlled by an on-delay timer consists of three parts, which

stand for (4)_____, (5)_____ and

(6)_____.

7-9.   The code that is added to a load controlled by an off-delay timer consists of three parts, which

stand for (7)_____, (8)_____ and

(9)_____.

10.   On a timer with an adjustable time setting of 0 to 10, representing 0 to 150 seconds, a setting of

2 would represent _____ seconds.

11.   The graph shown below illustrates a _____ timer.

```
CONTROL          ┌─────────┐
CIRCUIT   _____┘         └_____

 LOAD               ┌────┐
CIRCUIT   _____┘    └_____
```

12.   The graph shown below illustrates a _____ timer.

```
CONTROL      ┌────────┐
CIRCUIT   ___┘        └_____

 LOAD        ┌────────────────┐
CIRCUIT   ___┘                └_____
```

Continued

13. On the timer socket shown below, the coil voltage would be applied to pins

_____.

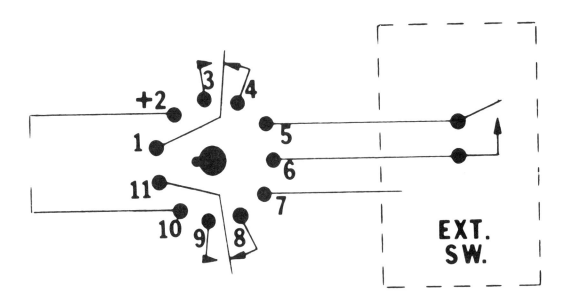

14-15. On the timer socket illustrated in Question #13, the NO contacts would be pins

(14)_____ and pins (15)_____.

16-17. On the timer socket illustrated in Question #13, the NC contacts would be pins

(16)_____ and pins (17)_____.

Continued

Follow the instructions in each statement.

18-19. Add the timing code (X and 0) for both of the loads illustrated in the following circuit.

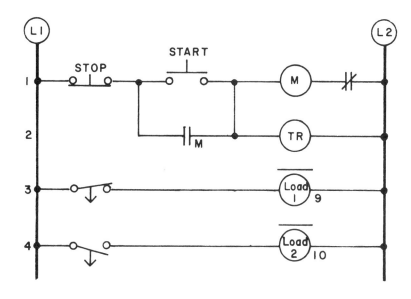

20-21. Add the timing code (X and O) for both of the loads illustrated in the following circuit.

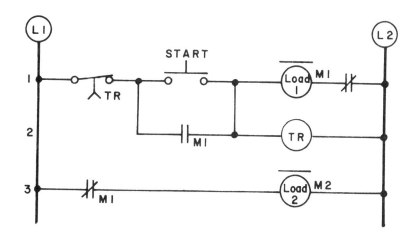

Continued

22-26. Complete the cross reference numbering for each of the controls.

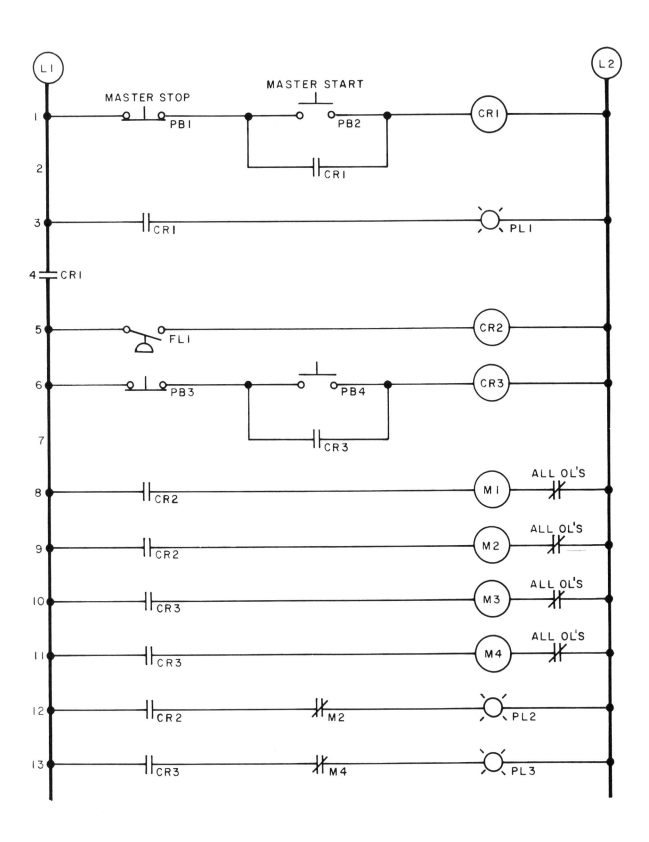

WORKSHEET 8-1

Score _____

Name _____ Class _____ Date _____

Complete the truth tables for the circuits shown below. Place an X in the appropriate box to indicate the correct answer.

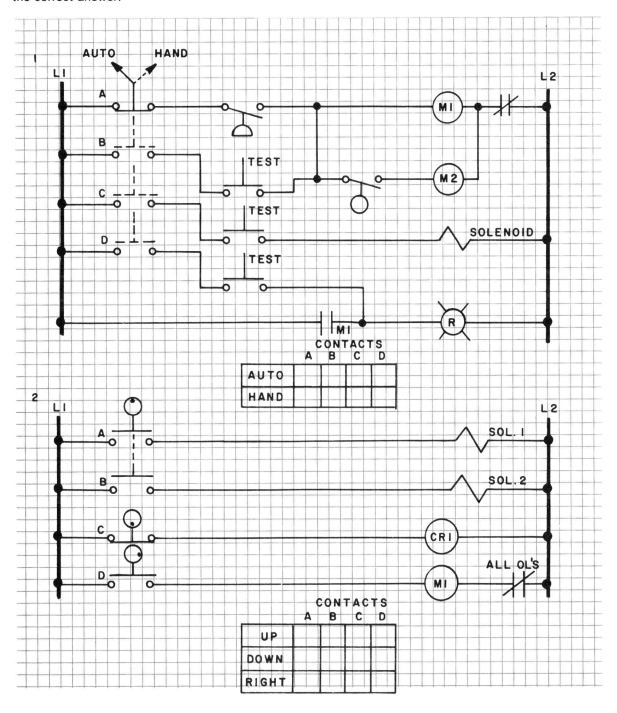

CONTACTS	A	B	C	D
AUTO				
HAND				

CONTACTS	A	B	C	D
UP				
DOWN				
RIGHT				

WORKSHEET 8-2

Name _____ Class _____ Date _____

Complete the line diagrams according to the circuit information given below. Use standard lettering, numbering, and coding information. Connecting lines should be straight and the circuits neatly drawn.

Circuit 1: Design a circuit in which a temperature switch is used to control a load even though the temperature switch contacts cannot directly handle the load.

Circuit 2: Complete the power circuit wiring diagram as a three-phase Wye-connected set of heating elements.

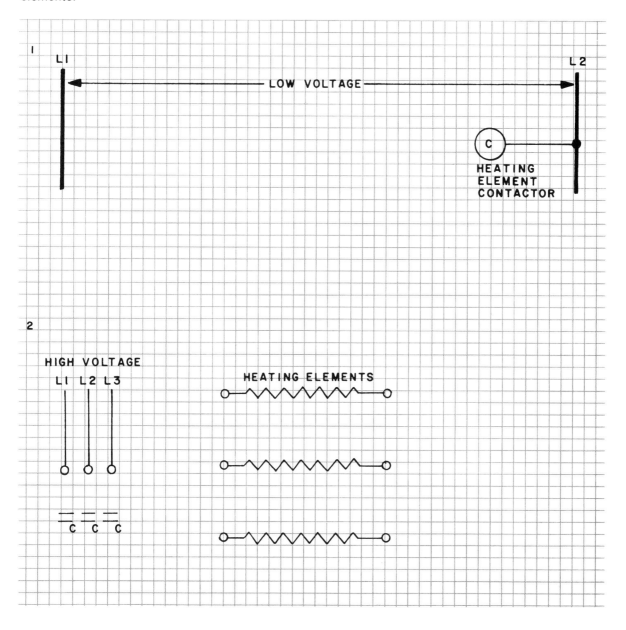

WORKSHEET 8-3

Score _____

Name _____ Class _____ Date _____

Complete the line diagrams according to the circuit information given below. Use standard lettering, numbering, and coding information. Connecting lines should be straight and the circuits neatly drawn.

Circuit 1: Design a circuit with a vacuum switch that sounds a warning bell in the case of a loss of vacuum.

Circuit 2: Redraw Circuit 1, adding a timer so that the vaccum switch does not activate the bell unless the loss of vacuum occurs after the pump has run for 60 seconds. This type of circuit will keep the bell from ringing until the pump has time to respond to the loss of vacuum.

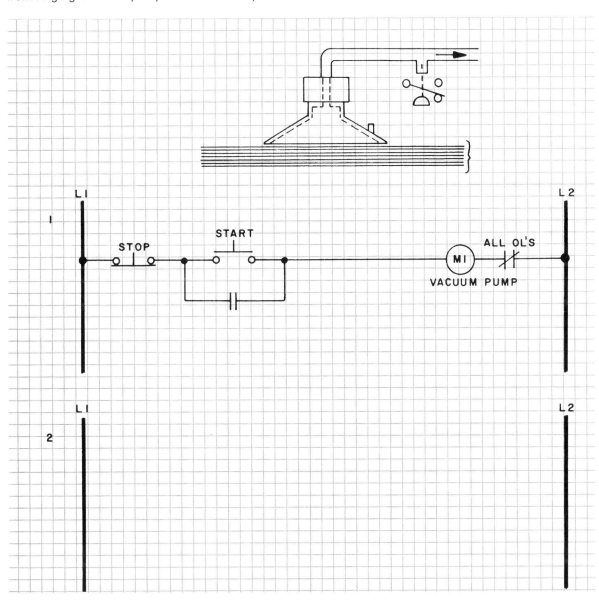

WORKSHEET 8-4

Score _____

Name _____ Class _____ Date _____

Complete the line diagram to design a circuit with three pressure switches to maintain the proper amount of air pressure in an inflatable building. One pressure switch should control an air pump to keep the building inflated. A second switch should detect overpressure that could rupture the building and should warn of it by sounding an alarm bell. The third switch should detect underpressure that could cause the building to collapse and should warn of it by sounding an alarm horn.

Use standard lettering, numbering, and coding information. Connecting lines should be straight and the circuit neatly drawn.

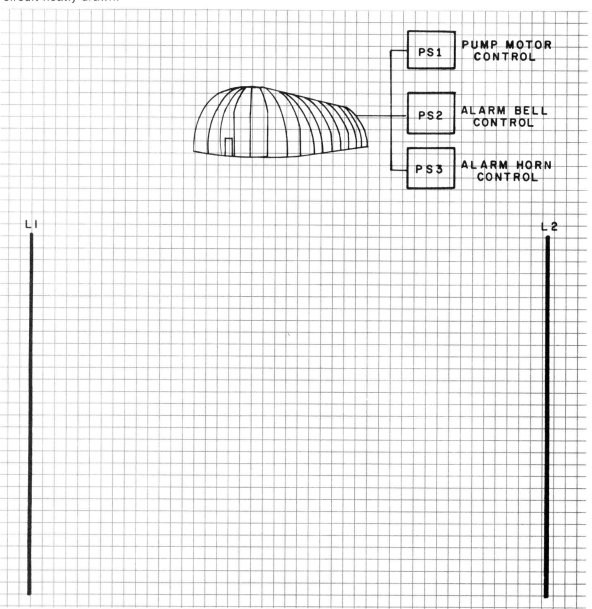

WORKSHEET 8-5

Score _____

Name _____ Class _____ Date _____

Complete the line diagram to design a circuit with two separate temperature switches and a selector switch to provide two temperature controls. Heat should be provided by heating elements activated through a magnetic contactor. The selector switch should have three settings: high, low, and off. Temperature Switch 1 should control the high temperature and Temperature Switch 2 should control the low temperature. Be sure you fill in the truth table for the selector switch to illustrate the circuit operation.

Use standard lettering, numbering, and coding information. Connecting lines should be straight and the circuit neatly drawn.

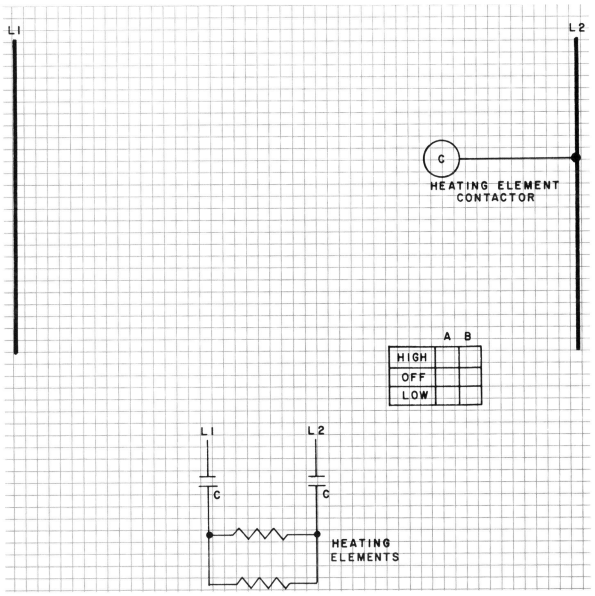

WORKSHEET 8-6

Score _____

Name _____ Class _____ Date _____

Complete the line diagram to design a circuit with a standard START-STOP pushbutton station to control a fan motor and an electric heater contactor. Add to this circuit a flow switch to ensure that the proper amount of air flow is present when the fan motor and heater are on. An alarm bell should sound if the flow is restricted when the fan is on, but it should not sound if the fan motor is off.

Use standard lettering, numbering, and coding information. Connecting lines should be straight and the circuit neatly drawn.

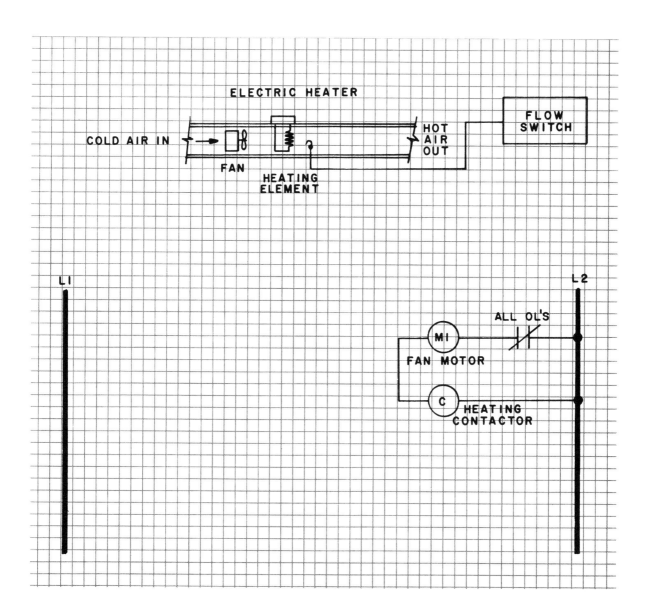

WORKSHEET 8-7

Score _____

Name _____ Class _____ Date _____

Complete the line diagrams according to the circuit information given below. Use standard lettering, numbering, and coding information. Connecting lines should be straight and the circuits neatly drawn.

Circuit 1: Redraw the circuit of Worksheet 8-6, adding a timer to keep the ventilation fan motor operating for 30 seconds after the heating element is turned off.

Circuit 2: Redraw the circuit of Worksheet 8-6, changing the circuit control from a pushbutton control to an automatic temperature control circuit.

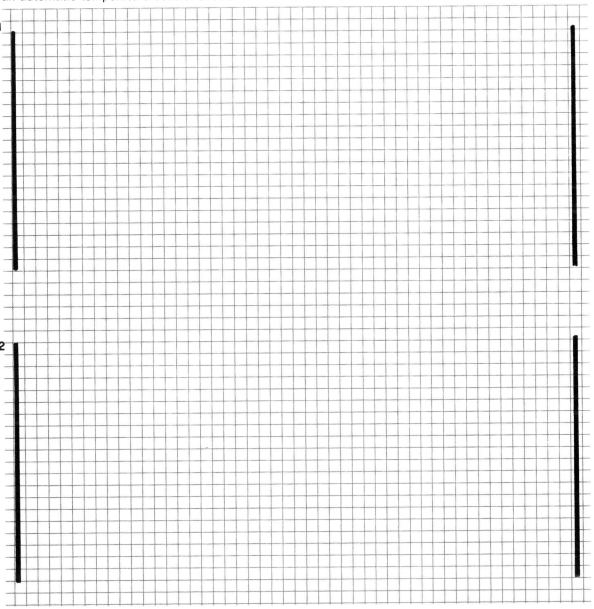

WORKSHEET 8-8

Name _____ Class _____ Date _____

Referring to Data Sheet F, and using the information on "level control for conductive liquids" (SV 115/215), design a circuit where this type of relay is connected to control a pump motor. The relay should be connected to monitor two levels of fluid. The pump motor should turn on when the maximum level setting is reached, and it should stay on until the minimum level setting is reached. Indicate in the line diagram only those connections necessary to form the control circuit. The wiring diagram shown is for information only and should not be connected into the line diagram.

Use standard lettering, numbering, and coding information. Connecting lines should be straight and the circuit neatly drawn.

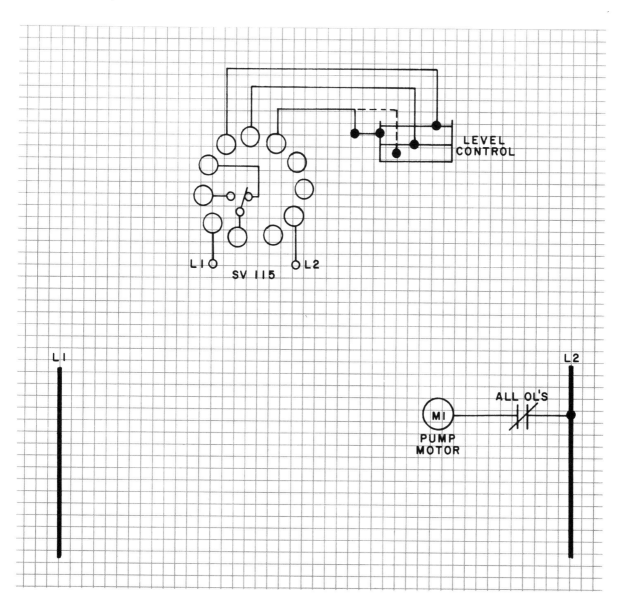

Applications and Installation of Control Devices

Select the best way to complete each statement and circle a, b, c, or d.

1. A pushbutton is a control device used to switch an electrical circuit
   a. manually.
   b. mechanically.
   c. automatically.

2. A pressure switch is a control device used to switch an electrical circuit
   a. manually.
   b. mechanically.
   c. automatically.

3. A temperature switch is a control device used to switch an electrical circuit
   a. manually.
   b. mechanically.
   c. automatically.

4. A flow switch is a control device used to switch an electrical circuit
   a. manually.
   b. mechanically.
   c. automatically.

5. A liquid level switch is a control device used to switch an electrical circuit
   a. manually.
   b. mechanically.
   c. automatically.

6. A limit switch is a control device used to switch an electrical circuit
   a. manually.
   b. mechanically.
   c. automatically.

7. In an electrical circuit the limit switch is connected between
   a. L2 and the load.
   b. L1 and the load.
   c. L1 and L2.
   d. two loads.

8. An actuator that can be shortened or formed to angles is the
   a. pushroller actuator.
   b. standard actuator.
   c. wobble-stick actuator.
   d. fork-lever actuator.

9. An actuator designed for applications where a direct thrust with a limited travel is accomplished is the
   a. pushroller actuator.
   b. standard actuator.
   c. wobble-stick actuator.
   d. fork-lever actuator.

Continued

10. An actuator used mostly with reciprocating movement, using maintained contacts, is the
    a. pushroller actuator.
    b. standard actuator.
    c. wobble-stick actuator.
    d. fork-lever actuator.

11. In an electrical circuit the temperature switch is connected between
    a. L2 and the load.
    b. L1 and the load.
    c. L1 and L2.
    d. two loads.

12. Accidental operation of a pushbutton is avoided with a
    a. mushroom head operator.
    b. illuminated operator.
    c. half-shrouded operator.
    d. flush operator.

13. A pushbutton and indicating light are combined into a single unit in the
    a. mushroom head operator.
    b. illuminated operator.
    c. half-shrouded operator.
    d. flush operator.

14. An operator that allows for easy emergency stops and operation with gloved hands is the
    a. mushroom head operator.
    b. illuminated operator.
    c. half-shrouded operator.
    d. flush operator.

15. The normally open contacts of a flow switch are "open" when there is
    a. no flow.
    b. flow in the pipe or duct.

16. In the circuit below, the limit switch is
    a. correctly installed.
    b. incorrectly installed.

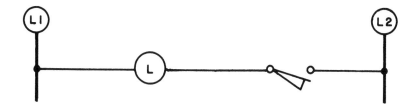

17. In the circuit below, the limit switch is
    a. correctly installed.
    b. incorrectly installed.

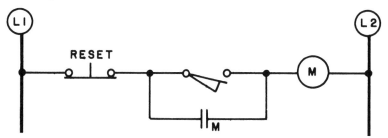

Continued

18. A liquid level switch or air pollution detector belongs to the
    a. signal part of the control circuit.
    b. decision part of the control circuit.
    c. action part of the control circuit.
    d. power part of the control circuit.

19. A logic module belongs to the
    a. signal part of the control circuit.
    b. decision part of the control circuit.
    c. action part of the control circuit.
    d. power part of the control circuit.

20. A motor starter belongs to the
    a. signal part of the control circuit.
    b. decision part of the control circuit.
    c. action part of the control circuit.
    d. power part of the control circuit.

Fill in the blanks to complete each statement.

21-22. Pushbutton contacts are available as either normally (21)_____ or

normally (22)_____.

23-25. Pressure switches respond to pressure from such media as (23)_____,

(24)_____ and (25)_____.

26. When using a pressure switch to maintain pressure in a tank, you should use the normally

_____ contacts.

27. The _____ setting is the pressure range between the rising pressure and

the falling pressure that is required to actuate the contacts.

28-29. The two main parts of a limit switch are the (28)_____ and

(29)_____.

30. When using a float switch for a sump operation, you should use the normally

_____ contacts.

31. When using a float switch for a pump operation, you should use the normally

_____ contacts.

Continued

79

32. The selector switch below has _____ positions.

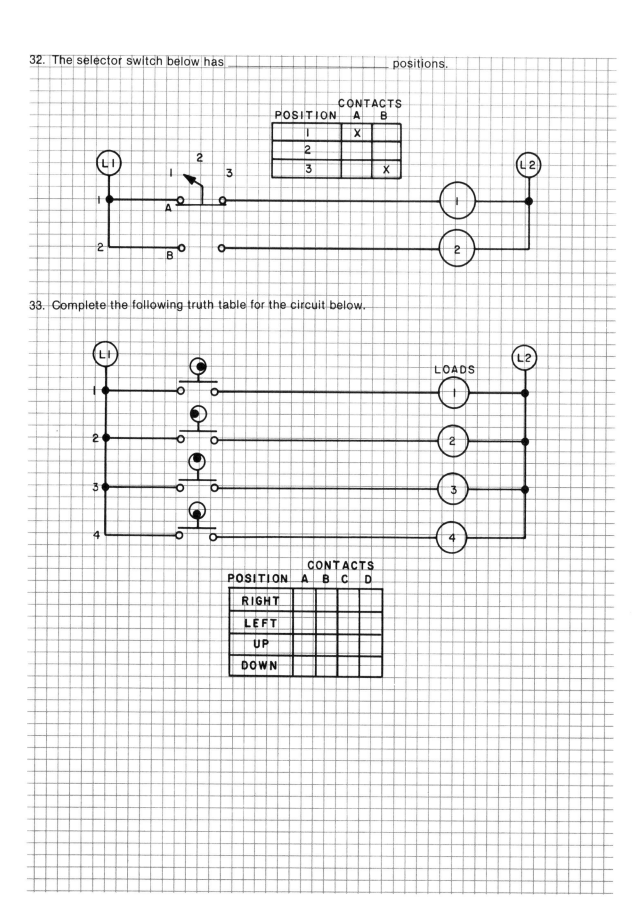

POSITION	CONTACTS	
	A	B
1	X	
2		
3		X

33. Complete the following truth table for the circuit below.

POSITION	CONTACTS			
	A	B	C	D
RIGHT				
LEFT				
UP				
DOWN				

WORKSHEET 9-1

Name _____ Class _____ Date _____

Referring to Data Sheet G, use the contacts below to draw a wiring diagram that will forward and reverse the motor at the voltage required. Assume that the wiring information on the motor nameplate for that voltage is: *To reverse rotation of motor, interchange leads T5 and T8.*

Although four sets of contacts are shown for each direction, use only the number of contacts required. Connecting lines should be straight and the circuit neatly drawn.

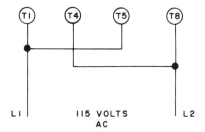

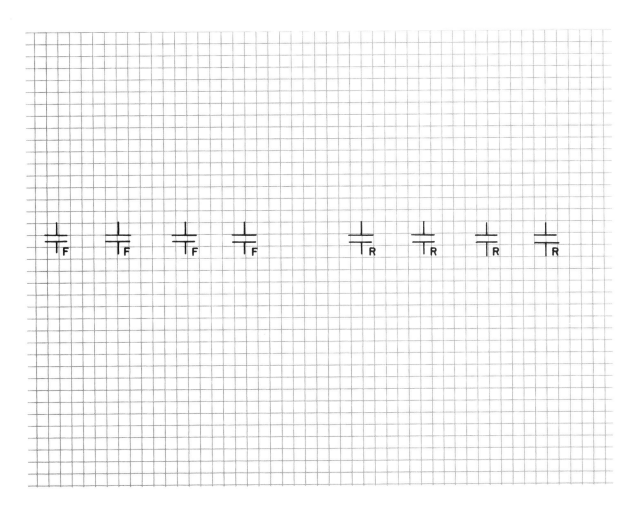

WORKSHEET 9-2

Name _____ Class _____ Date _____

Referring to Data Sheet G, use the contacts below to draw a wiring diagram that will forward and reverse the motor at the voltage required. Assume that the wiring information on the motor nameplate for that voltage is: *To reverse rotation of motor, interchange any two power leads.* The accepted standard is to interchange leads T1 and T3.

Although four sets of contacts are shown for each direction, use only the number of contacts required. Connecting lines should be straight and the circuit neatly drawn.

ILLUSTRATE WIRING DIAGRAM FOR 230 VOLTS

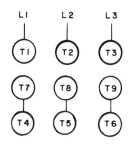

460 VOLTS
AC

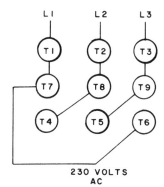

230 VOLTS
AC

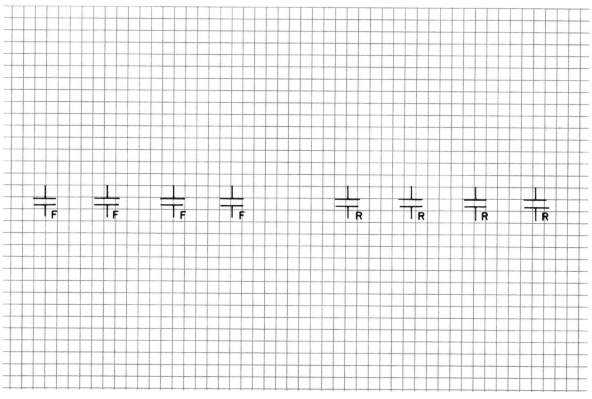

WORKSHEET 9-3

Score _____

Name _____ Class _____ Date _____

Referring to Data Sheet G, use the contacts below to draw a wiring diagram that will forward and reverse the motor at the voltage required. Assume that the wiring information on the motor nameplate for that voltage is: *To reverse rotation of motor, interchange the red and black lead.*

Although four sets of contacts are shown for each direction, use only the number of contacts required. Connecting lines should be straight and the circuit neatly drawn.

ILLUSTRATE WIRING DIAGRAM FOR 120 VOLTS

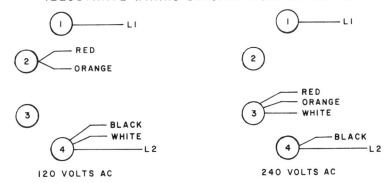

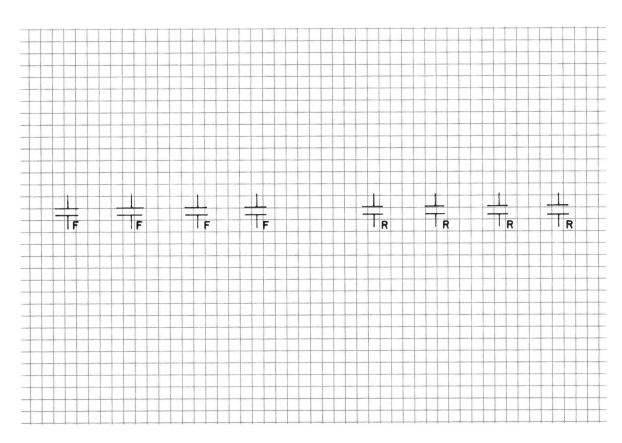

WORKSHEET 9-4

Name _____ Class _____ Date _____

Referring to Data Sheet G, use the contacts below to draw a wiring diagram that will forward and reverse the motor at the voltage required. Assume that the wiring information on the motor nameplate for that voltage is: *To reverse rotation of motor, interchange the red and black lead.*

Although four sets of contacts are shown for each direction, use only the number of contacts required. Connecting lines should be straight and the circuit neatly drawn.

ILLUSTRATE WIRING DIAGRAM FOR 230 VOLTS

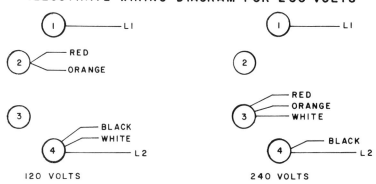

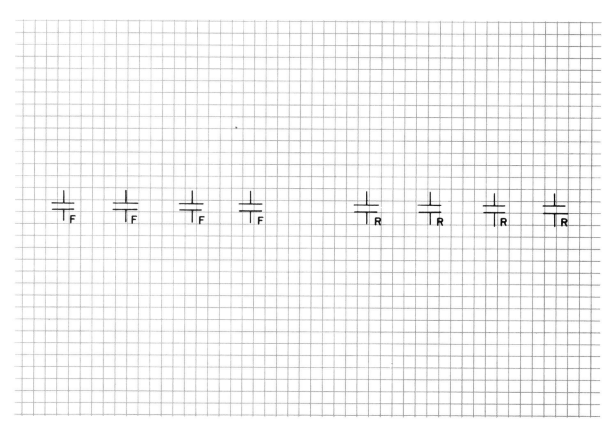

WORKSHEET 9-5

Score _____

Name _____ Class _____ Date _____

Referring to Data Sheet G, use the contacts below to draw a wiring diagram that will forward and reverse the motor at the voltage required. Assume that the wiring information on the motor nameplate for that voltage is: *To reverse rotation of motor, interchange the red and black lead.*

Although four sets of contacts are shown for each direction, use only the number of contacts required. Connecting lines should be straight and the circuit neatly drawn.

ILLUSTRATE WIRING DIAGRAM FOR 230 VOLTS

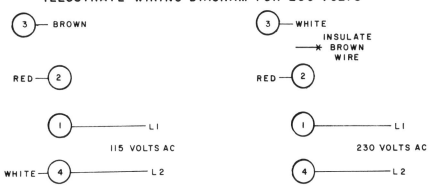

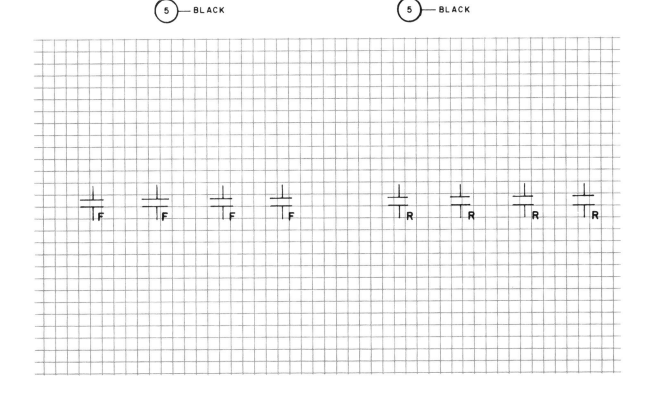

WORKSHEET 9-6

Score _____

Name _____ Class _____ Date _____

Referring to Data Sheet G, use the contacts below to draw a wiring diagram that will forward and reverse the motor at the voltage required. Assume that the wiring information on the motor nameplate for that voltage is: *To reverse rotation of motor, interchange the red and black lead.*

Although four sets of contacts are shown for each direction, use only the number of contacts required. Connecting lines should be straight and the circuit neatly drawn.

ILLUSTRATE WIRING DIAGRAM FOR 115 VOLTS

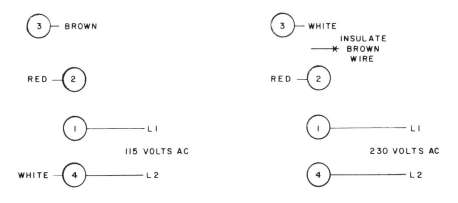

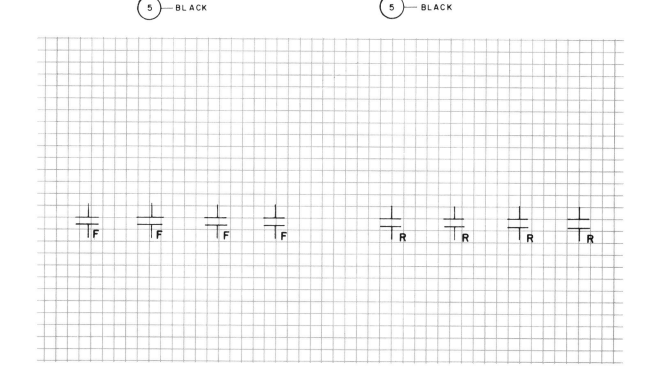

WORKSHEET 9-7

Name _____ Class _____ Date _____

Referring to Data Sheet G, use the contacts below to draw a wiring diagram that will forward and reverse the motor at the voltage required. Assume that the wiring information on the motor nameplate for that voltage is: *To reverse rotation of motor, interchange the red and black lead.*

Although four sets of contacts are shown for each direction, use only the number of contacts required. Connecting lines should be straight and the circuit neatly drawn.

ILLUSTRATE WIRING DIAGRAM FOR LOW SPEED

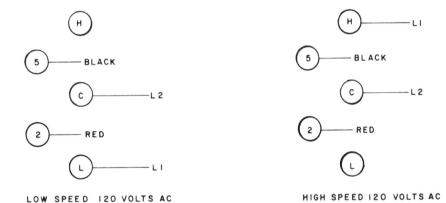

LOW SPEED 120 VOLTS AC   HIGH SPEED 120 VOLTS AC

WORKSHEET 9-8

Name _____ Class _____ Date _____

Referring to Data Sheet G, use the contacts below to draw a wiring diagram that will forward and reverse the motor at the voltage required. Assume that the wiring information on the motor nameplate for that voltage is: *To reverse rotation of motor, interchange the red and black lead.*

Although four sets of contacts are shown for each direction, use only the number of contacts required. Connecting lines should be straight and the circuit neatly drawn.

ILLUSTRATE WIRING DIAGRAM FOR HIGH SPEED

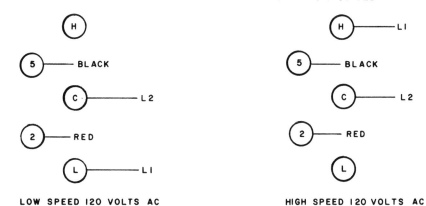

LOW SPEED 120 VOLTS AC   HIGH SPEED 120 VOLTS AC

WORKSHEET 9-9

Name _____ Class _____ Date _____

Complete the wiring diagram on the other side of the page according to the line diagram shown below of a standard FORWARD-REVERSE-STOP pushbutton station for forwarding and reversing a motor. Included in this circuit are both mechanical and auxiliary contact interlocking. Also included are a forward limit switch to stop the motor in forward and a reverse limit switch to stop the motor in reverse. Overload protection is common to both forward and reverse.

Your connecting lines should be straight and the circuit neatly drawn. Do not make any wire splices or additional terminal connections on the wiring diagram. All connections must run from terminal screw to terminal screw.

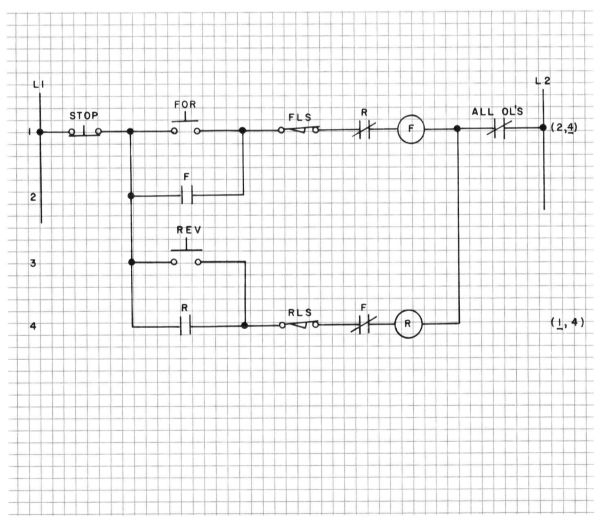

Continued

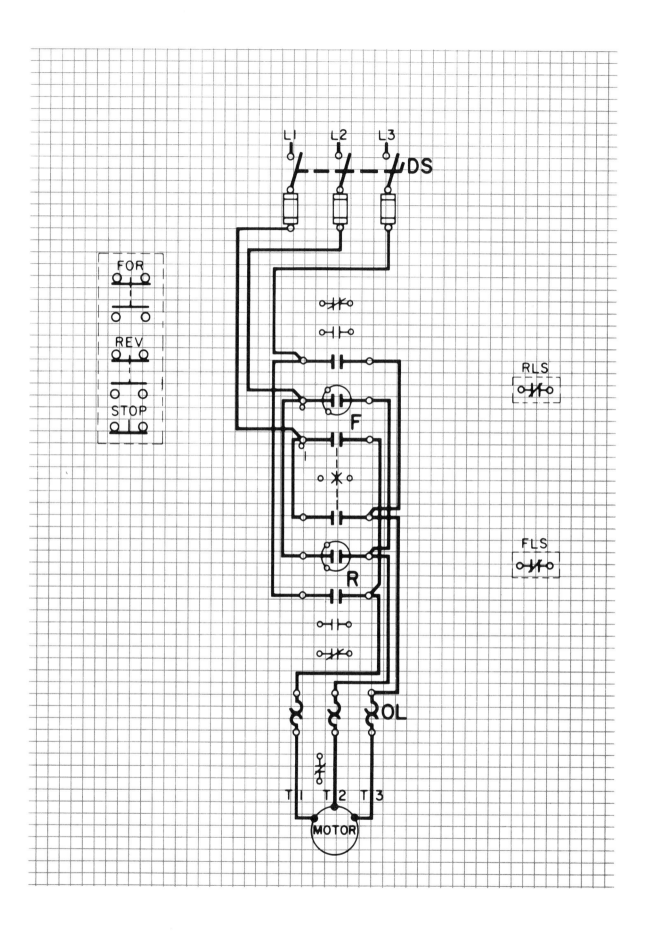

WORKSHEET 9-10

Name _____ Class _____ Date _____

Complete the wiring diagram on the other side of the page according to the line diagram shown below of a FORWARD-REVERSE-STOP pushbutton station with indicating lights to show the direction of motor travel. These lights are to be mounted within the pushbutton enclosure. Overload protection is common to both forward and reverse.

Your connecting lines should be straight and the circuit neatly drawn. Do not make any wire splices or additional terminal connections on the wiring diagram. All connections must run from terminal screw to terminal screw.

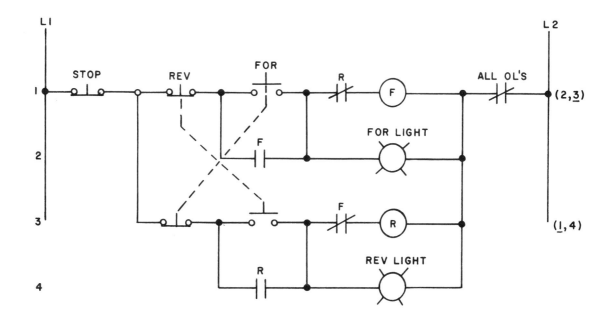

Continued

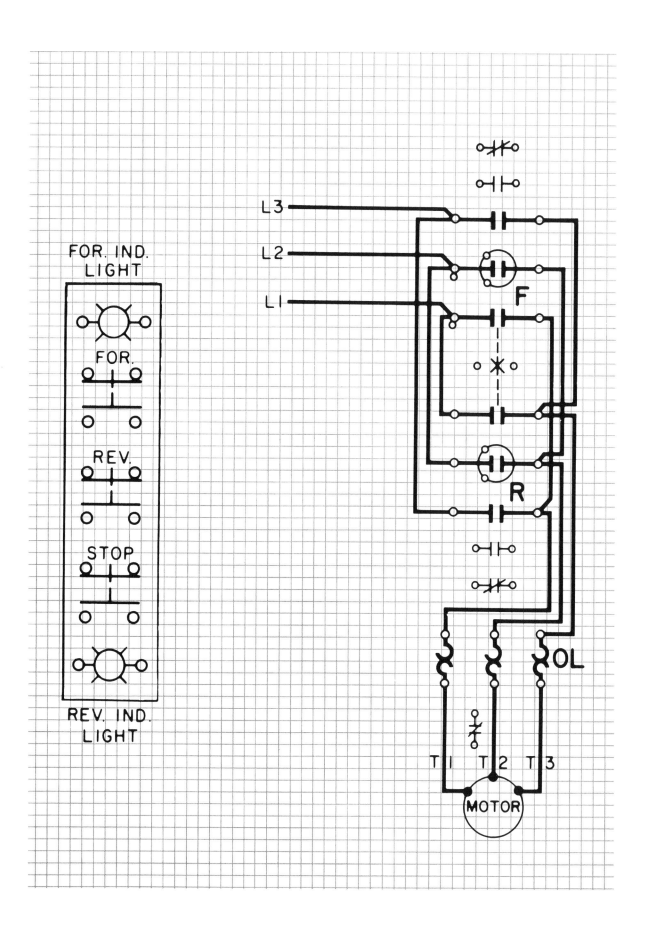

WORKSHEET 9-11

Score _____

Name _____ Class _____ Date _____

Complete the wiring diagram on the other side of the page according to the line diagram shown below of a standard START-STOP station with selector switch to control direction of motor travel. In case the motor and drive unit cannot be seen from the control station, a visual indication of the direction of travel is provided on the selector switch. Overload protection is common to forward and reverse.

Your connecting lines should be straight and the circuit neatly drawn. Do not make any wire splices or additional terminal connections on the wiring diagram. All connections must run from terminal screw to terminal screw.

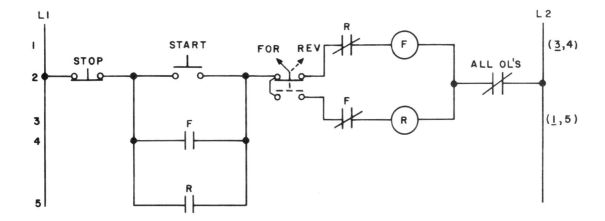

Continued

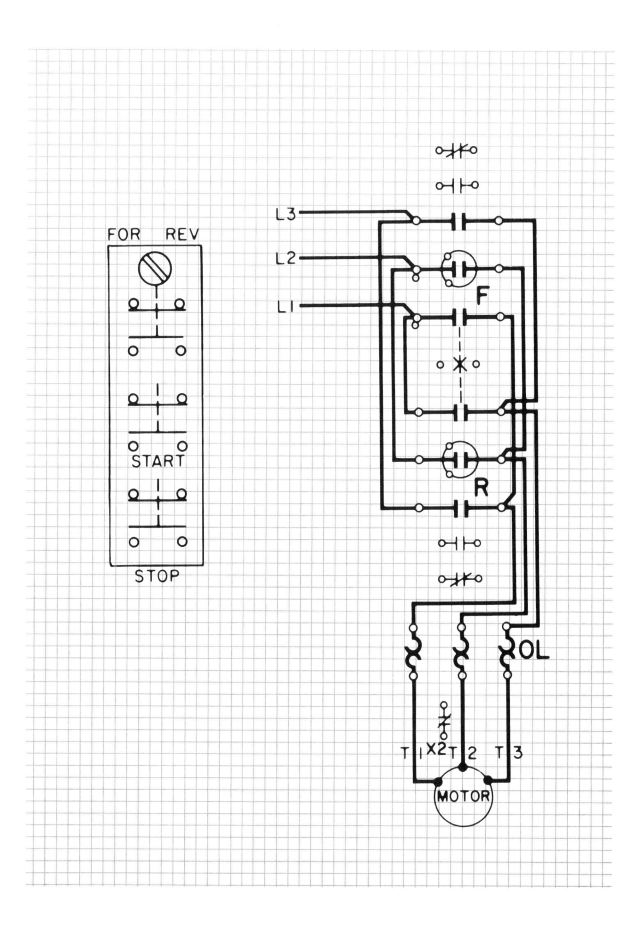

WORKSHEET 9-12

Name _____ Class _____ Date _____

Complete the wiring diagram on the other side of the page according to the line diagram shown below of a standard FORWARD-REVERSE-STOP pushbutton station with a selector switch to provide for jogging or running. When the selector switch is in the JOG position, the FORWARD and REVERSE pushbuttons will energize the motor only as long as they are pressed. When the selector switch is in the RUN position, the FORWARD and REVERSE pushbuttons will operate as a standard pushbutton station with MEMORY. Overload protection is common to both forward and reverse.

Your connecting lines should be straight and the circuit neatly drawn. Do not make any wire splices or additional terminal connections on the wiring diagram. All connections must run from terminal screw to terminal screw.

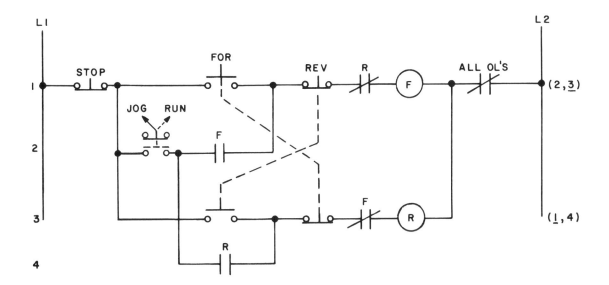

Continued

95

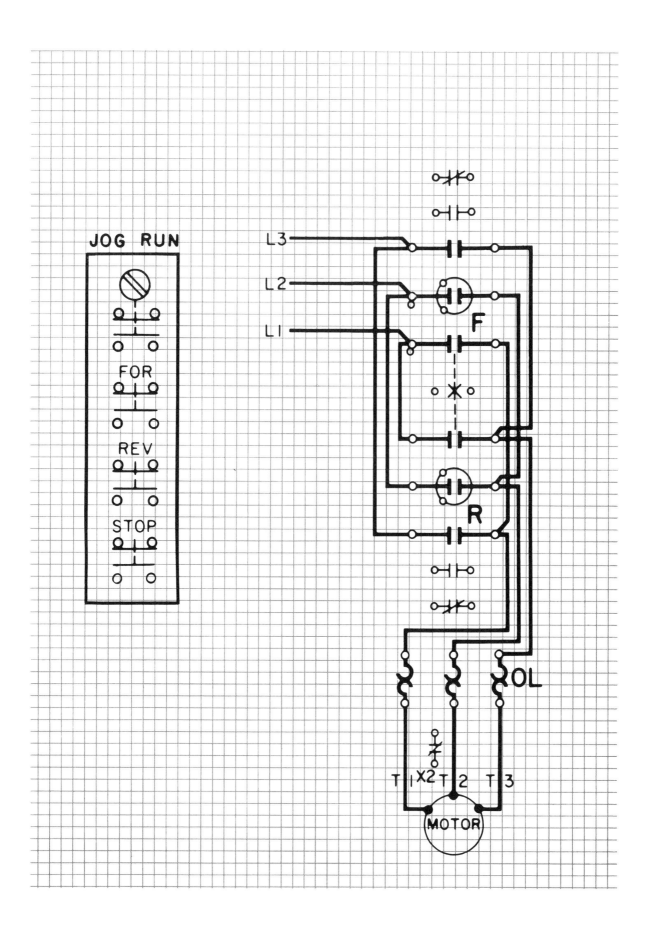

JOG RUN

FOR

REV

STOP

L3

L2

L1

F

R

OL

T1 X2 T2 T3

MOTOR

Name _____

Class _____ Date _____

Score _____

**Reversing Circuits Applied to Single-Phase,
Three-Phase and DC Motor Types**

Follow the instructions in each statement.

1. Mark each of the nine motor leads (T1, T2, etc.) for the motor below.

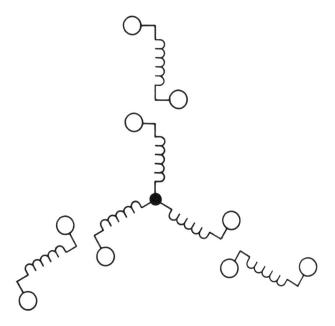

2. Mark each of the nine motor leads (T1, T2, etc.) for the motor below.

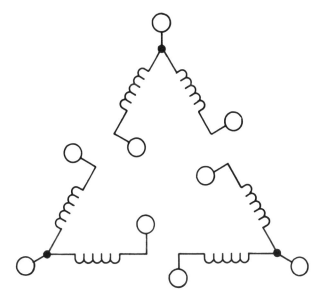

Continued

97

3. Mark each of the motor leads for the motor below.

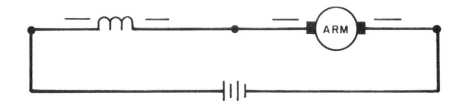

4. Mark each of the motor leads for the motor below.

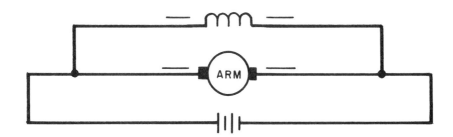

5. Mark each of the motor leads for the motor below.

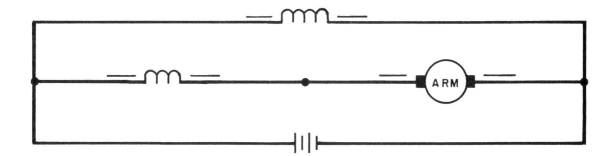

Continued

98

Fill in the blanks to complete each statement.

6. The motor illustrated in Question #1 is a _____ motor.

7. The motor illustrated in Question #2 is a _____ motor.

8. The motor illustrated in Question #3 is a _____ motor.

9. The motor illustrated in Question #4 is a _____ motor.

10. The motor illustrated in Question #5 is a _____ motor.

11. Although any two lines can be interchanged to reverse a three-phase motor, the industrial standard is to interchange _____.

12. The device used to disconnect the starting winding of a single-phase motor is called the _____.

13. The accepted industrial standard is to interchange the leads of the _____ winding to reverse a singe-phase motor.

14. In using an ohmmeter to determine the starting and running windings of a motor, it is important to remember that the _____ winding usually has the greatest resistance.

15. To develop more starting torque, a _____ is added to the single-phase motor.

16-18. The three types of interlocking methods used in reversing circuits and controls are

(16)_____, (17)_____ and

(18)_____.

19. A _____ circuit allows the operator to start the motor for a short time without MEMORY.

20. A drum switch is not considered to be a motor starter because the switch does not contain _____.

WORKSHEET 10-1

Name _____ Class _____ Date _____

Complete the transformer wiring diagrams according to the Delta-Delta transformer bank connections shown below. Indicate how the primary transformer lines will be connected to the distribution system to form a Delta primary. Indicate how the secondary transformer lines will be connected to the distribution systems that will provide three-phase, high voltage single-phase, and low voltage single-phase power.

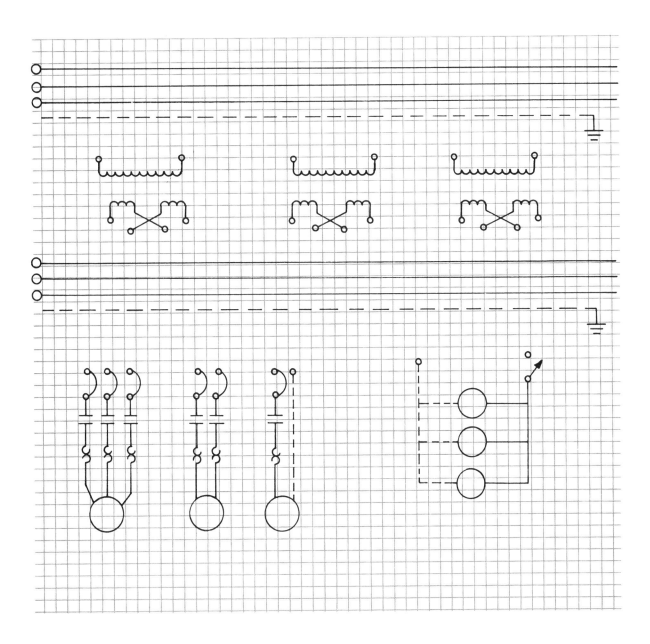

WORKSHEET 10-2

Score _____

Name _____ Class _____ Date _____

Complete the transformer wiring diagrams according to the Wye-Wye transformer bank connections shown below. Indicate how the primary transformer lines will be connected to the distribution system to form a Wye primary. Indicate how the secondary transformer lines will be connected to the distribution system to form a Wye secondary that will provide three-phase, high voltage single-phase, and low voltage single-phase power.

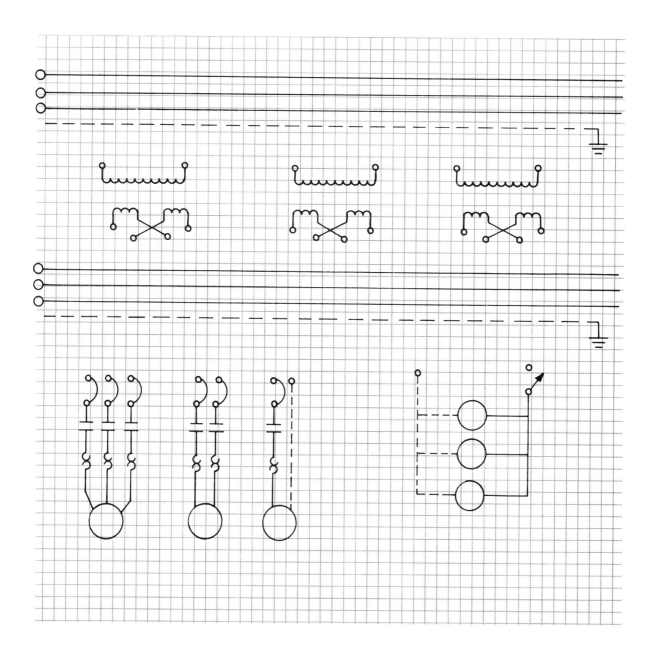

WORKSHEET 10-3

Name _____   Class _____   Date _____

Complete the transformer wiring diagrams according to the Delta-Wye transformer bank connections shown below. Indicate how the primary transformer lines will be connected to the distribution system to form a Delta primary. Indicate how the secondary transformer lines will be connected to the distribution system to form a Wye secondary that will provide three-phase, high voltage single-phase, and low voltage single-phase power.

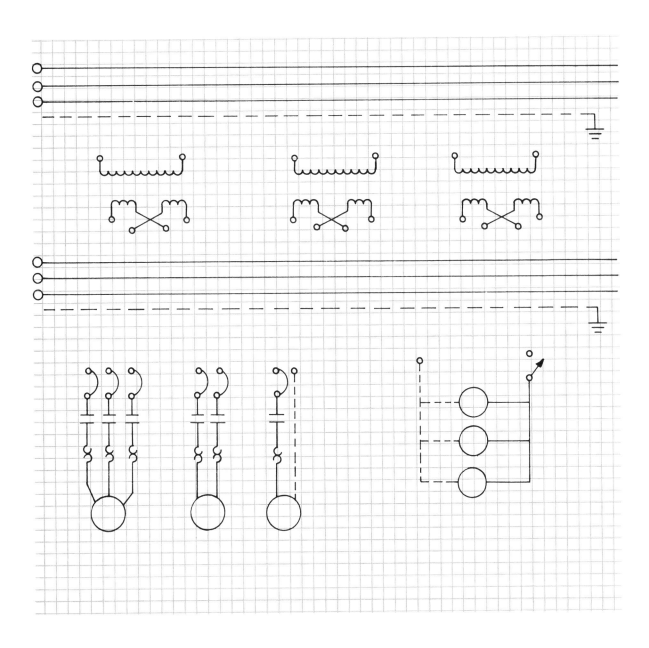

WORKSHEET 10-4

Name _____   Class _____   Date _____

Complete the transformer wiring diagrams according to the Wye-Delta transformer bank connections shown below. Indicate how the primary transformer lines will be connected to the distribution system to form a Delta secondary that will provide three-phase, high voltage single-phase, and low voltage single-phase power.

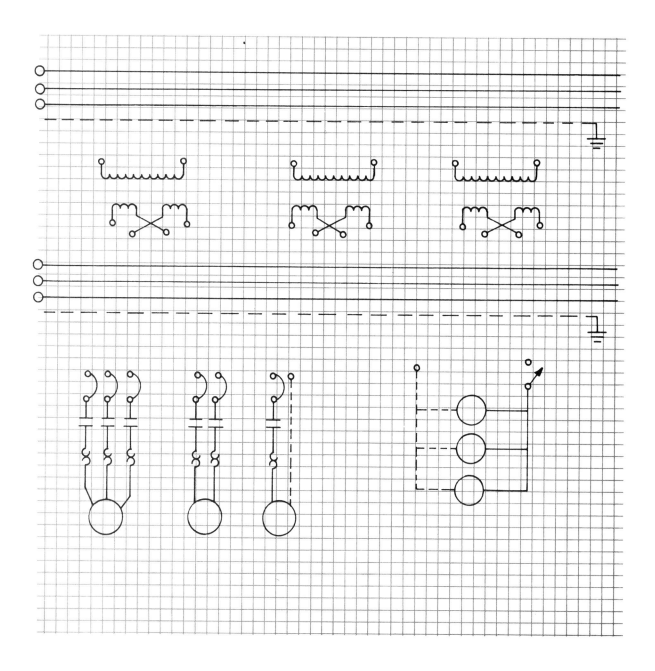

WORKSHEET 10-5

Name _____   Class _____   Date _____

Referring to Data Sheet H, make a simple sketch of the busway system shown below. On the other side of the page fill in the bill of materials based on pricing information and dimensions.

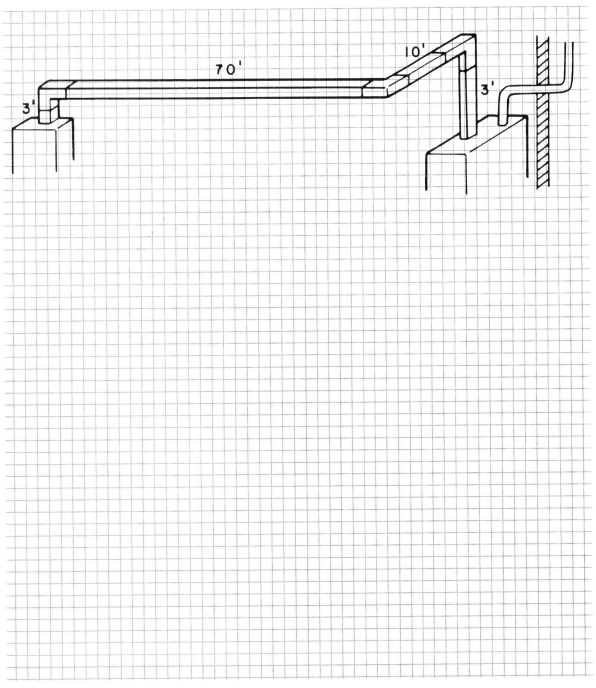

Continued

PRICING INFORMATION (LIST)

Feeder busway duct .... $ 40 per foot
Plug-in busway duct .... 45 per foot
Elbows..................... 170 each

Crosses .... $225 each
50 amp breakers ..... 175 each
100 amp breakers .... 225 each

BILL OF MATERIALS

Totals

_____ feet of feeder duct @ $_____ per foot $_____

_____ feet of plug-in duct @ $_____ per foot   _____

_____ elbows @ $_____ each   _____

_____ tees @ $_____ each   _____

_____ crosses @ $_____ each   _____

_____ 50 amp breakers @ $_____ each   _____

_____ 100 amp breakers @ $_____ each   _____

GRAND TOTAL _____ (LIST)

− 20% DISCOUNT _____

TOTAL _____

WORKSHEET 10-6

Name _____   Class _____   Date _____

Based on the following pricing information, and on the dimensions in the illustration below, fill in the bill of materials on the other side of the page.

PRICING INFORMATION (LIST)

Feeder busway duct	$ 40 per foot	Crosses	$225 each
Plug-in busway duct	45 per foot	50 amp breakers	175 each
Elbows......................	170 each	100 amp breakers	225 each
Tees	190 each		

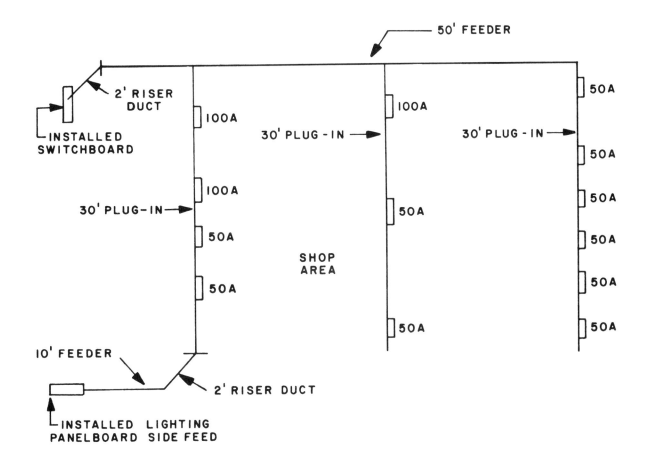

Continued

107

BILL OF MATERIALS

Totals

_____ feet of feeder duct @ $_____ per foot $_____

_____ feet of plug-in duct @ $_____ per foot   _____

_____ elbows @ $_____ each   _____

_____ tees @ $_____ each   _____

_____ crosses @ $_____ each   _____

_____ 50 amp breakers @ $_____ each   _____

_____ 100 amp breakers @ $_____ each   _____

GRAND TOTAL _____

− 20% DISCOUNT _____

TOTAL _____

WORKSHEET 10-7

Name _____ Class _____ Date _____

Fill in the current, voltage, and phase rating for each of the receptable configurations shown below.

1   (◯)   E ____   I ____   Ø ____

6   (◯)   E ____   I ____   Ø ____

2   (◯)   E ____   I ____   Ø ____

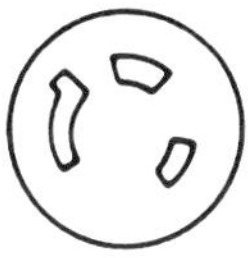

7   (◯)   E ____   I ____   Ø ____

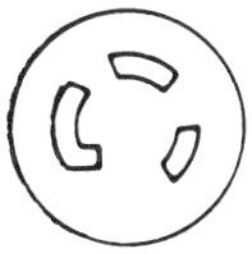

3   (◯)   E ____   I ____   Ø ____

8   (◯)   E ____   I ____   Ø ____

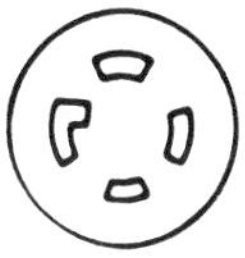

4   (◯)   E ____   I ____   Ø ____

9   (◯)   E ____   I ____   Ø ____

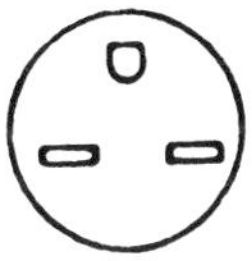

5   (◯)   E ____   I ____   Ø ____

10   (◯)   E ____   I ____   Ø ____

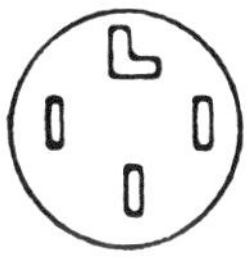

WORKSHEET 10-8

Score _____

Name _____ Class _____ Date _____

Complete the wiring diagram according to the line diagram shown below. The blank motor control panel is to be wired as a standard START-STOP pushbutton control with MEMORY and an indicating light. Connect the wiring of the motor control panel to the terminal blocks provided. Do not connect the pushbutton or indicator light directly to the terminal blocks. Use blank spaces adjoining the pushbutton and indicator light to indicate by number which terminals would be connected to these points. For example, the stop pushbutton would be marked with wire number 1 and 2.

Your connecting lines should be straight and the circuit neatly drawn. Do not make any wire splices or additional terminal connections on the wiring diagram. All connections must run from terminal screw to terminal screw.

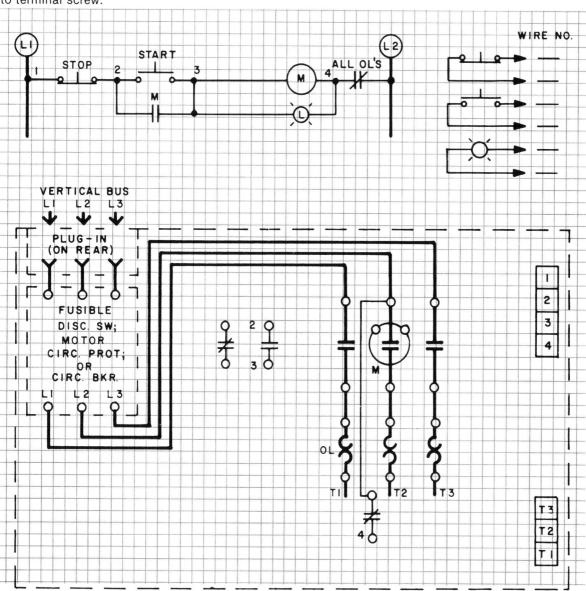

WORKSHEET 10-9

Name _____   Class _____   Date _____

Complete the wiring diagram according to the line diagram shown below. The blank motor control panel is to be wired as a jogging circuit with a control relay. Connect the wiring of the motor control panel to the terminal blocks provided. Do not connect control devices directly to the terminal blocks. Use blank spaces adjoining the control devices to indicate by number which terminals would be connected to these points.

Your connecting lines should be straight and the circuit neatly drawn. Do not make any wire splices or additional terminal connections on the wiring diagram. All connections must run from terminal screw to terminal screw.

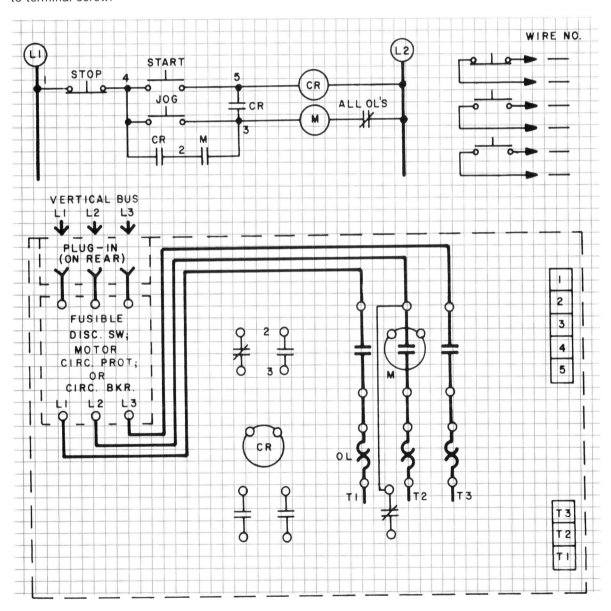

WORKSHEET 10-10

Name _____   Class _____   Date _____

Complete the wiring diagram according to the line diagram shown below. The blank motor control panel is to be wired as a circuit that provides surge and backspin protection by means of a time-delay relay. Do not connect control devices directly to the terminal blocks. Use blank spaces adjoining control devices to indicate by number which terminals would be connected to these points.

Your connecting lines should be straight and the circuit neatly drawn. Do not make any wire splices or additional terminal connections on the wiring diagram. All connections must run from terminal screw to terminal screw.

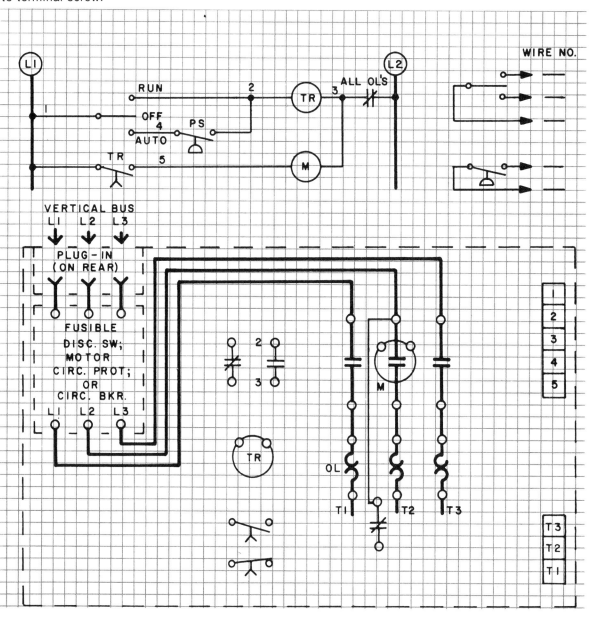

TECH-CHEK ✔ 10

Name _____

Class _____ Date _____

Score _____

**Power Distribution Systems, Transformers,
Switch boards, Panel boards, Motor Control Centers
and Busways**

Fill in the blanks to complete each statement.

1. Although there are several minor sources of electrical energy, the major source of

   electrical power is the _____.

2. When the three ends of the separate phases are connected together making a common wire,

   the connections are called a _____ connection.

3. When the end of one separate phase is connected to the beginning of the next phase, and so

   on, the connections are called a _____ connection.

4-5. In a transformer the (4)_____ winding is the coil that draws power from

   the source. The (5)_____ winding is the coil that delivers the energy at a

   transformed or changed voltage to the load.

6. The term "step-up" or "step-down" when used with transformers always refers to the

   _____.

7. In transformers a marking of X1 or X2 indicates the _____ (higher or

   lower) voltage side.

8. To provide for a variable output, the transformer is provided with _____.

9. As far as the power company is concerned, the _____ is the last point on

   the power distribution system.

10-11. Switchboards are designed for use as (10)_____,

   (11)_____, or a combination of both.

12-14. The three types of panelboards are the (12)_____,

   (13)_____ and (14)_____ panels.

15. The portion of the distribution system between the final overcurrent device protecting the

   circuit and the outlet or load connected to them is called the _____

   circuit.

16-17. The two basic types of busway ducts are the (16)_____ and

   (17)_____.

Continued

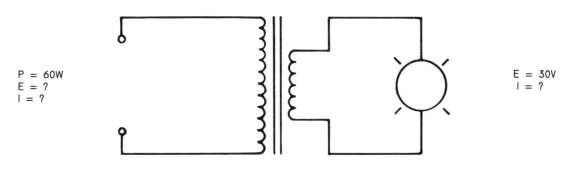

P = 60W
E = ?
I = ?

E = 30V
I = ?

TRANSFORMER RATIO = 4 TO 1

Use the information given in the transformer circuit above to answer questions 18-20.

18. What is the voltage on the primary side? _____

19. What is the current on the primary side? _____

20. What is the current on the secondary side? _____

WORKSHEET 11-1

Name _____ Class _____ Date _____

Identify the following parts of a PC board and the components commonly mounted on a PC board.

_____ 1. Insulated board

_____ 2. Traces

_____ 3. Terminal contacts

_____ 4. Pads

_____ 5. Bus

_____ 6. Edge card connector

_____ 7. Components

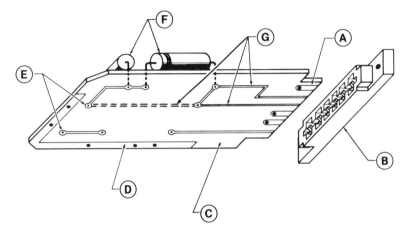

_____ 8. Resistors

_____ 9. Wafer switch

_____10. Power transformer

_____11. Transistor

_____12. SCR

_____13. Mini-dip IC

_____14. T0-5 IC

_____15. Trimmer resistor

_____16. Capacitor

_____17. Dual-in-line IC

_____18. Large scale IC

_____19. Heat sink

_____20. Dip switch

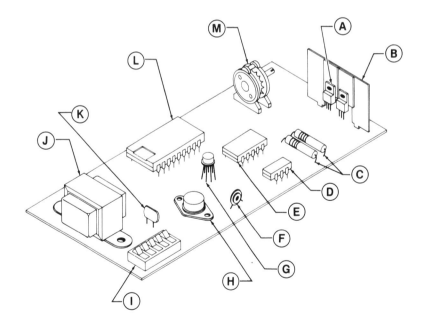

WORKSHEET 11-2

Name _____ Class _____ Date _____

1. Atoms contain three fundamental particles called _____,

   _____, and _____ .

2. Protons carry a _____ charge.

3. _____ carry no electrical charge.

4. Electrons are _____ charged.

5. _____ _____ determine the conductive or insulative

   value of a given material.

6. _____ is the process by which the crystal structure is altered.

7. The two types of material created by the addition of new atoms into a crystal

   are _____ and _____ .

8. _____ flow is equal to and opposite of electron flow.

9. _____ are components that have the unique ability of allowing current to

   pass through them in only one direction.

Identify the operation of a diode.

FORWARD BIAS
CATHODE — ANODE
CURRENT FLOW
HEAT SINK

FORWARD CURRENT (mA)

E

500 400 300 200 100

0.3 V   0.6 V   0.9 V
DEPLETION VOLTAGE

REVERSE CURRENT (µA)

A

D

C

B

AVALANCHE CURRENT
SMOKE

CURRENT FLOW

REVERSE BIAS

NOTE CHANGE OF SCALE FROM REVERSE TO FORWARD

_____10. Forward bias voltage

_____11. Reverse breakdown

_____12. Reverse bias voltage

_____13. Reverse current

_____14. Forward operating current

15. The maximum reverse bias voltage a diode can withstand is _____ .
    a. reverse bias
    b. peak inverse voltage
    c. avalanche current
    d. source voltage

WORKSHEET 11-3

Name _____   Class _____   Date _____

1. _____ is the changing of AC into DC.

2. List three diode rectifiers commonly used.

   a. _____   b. _____   c. _____

3. Indicate the direction of current flow through the full-wave bridge rectifiers for sequence 1-2-3.

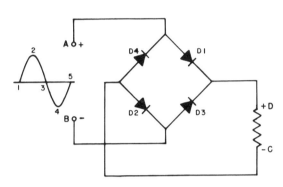

SEQUENCE 1-2-3: A is positive (+) and B is negative (−) with respect to A. D1 and D2 are foward biased and conduct. During conduction there is about 0.7V drop across D1 and D2 with current flow as indicated by the arrows. D3 and D4 are reversed biased and do not conduct.

4. The _____ _____ acts as a voltage regulator either by itself or in conjunction with other semiconductor devices.

5. _____ direct current eliminates pulsations and provides direct current at a constant intensity.

Identify the operating characteristics of a zener diode.

_____ 6. Zener breakdown

_____ 7. Forward breakover voltage

_____ 8. Forward current

_____ 9. Standard diode operating range

_____ 10. Reverse bias voltage

_____ 11. Forward bias voltage

_____ 12. Zener with operating range

_____ 13. Reverse current

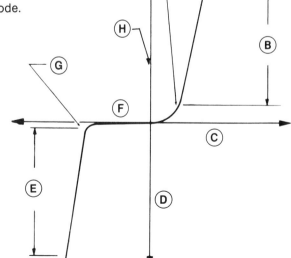

14. A(n) _____ _____ rectifier circuit uses three rectifier diodes to convert AC to DC.

117

WORKSHEET 11-4

Name _____ Class _____ Date _____

_____1. Thermistor

_____2. Photoconductive cell

_____3. Photovoltaic cell

_____4. Photoconductive diode

_____5. Hall effect sensor

_____6. Solid state pressure sensor

_____7. Light emitting diode

a. Responds to magnetic influence

b. Converts solar energy to electrical energy

c. A thermally sensitive resistor

d. Changes resistance with pressure

e. A light-sensitive resistor

f. A light-sensitive diode

g. Emits light when forward biased

8. Identify the following solid state electronic control devices.

a. _____

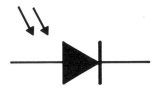

d. _____

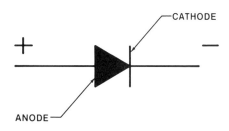

b. _____

e. _____

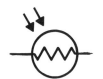

c. _____

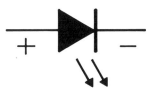

f. _____

WORKSHEET 11-5

Name _____ Class _____ Date _____

1. The two types of transistors are the _____ and the _____ .

2. Identify the parts of the transistor.

   a. _____ _____

   b. _____

   c. _____

3. In any transistor circuit the _____ - _____ junction

   must always be forward-biased and the base-collector junction must always be reverse-biased.

4. The _____ _____ I_B is the critical factor in determining the

   amount of current flow in a transistor.

5. A transistor switched ON is usually operating in the _____ region.

6. When a transistor is switched OFF, it is operating in the _____ region.

7. _____ is the ratio of the amplitude of the output signal to the amplitude of

   the input signal.

8. _____ amplifiers are two or more amplifiers used to obtain additional gain.

Complete the following diagrams, indicating where the load would be placed in a common-emitter and common-base amplifier.

9. **COMMON-EMITTER**

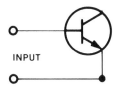

INPUT

10. **COMMON-BASE**

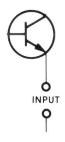

INPUT

WORKSHEET 11-6

Score _____

Name _____ Class _____ Date _____

Identify the operation of an SCR.

_____1. Forward breakover voltage

_____2. Holding current

_____3. Reverse breakdown voltage

_____4. Avalanche current

_____5. Reverse current

_____6. Forward blocking current

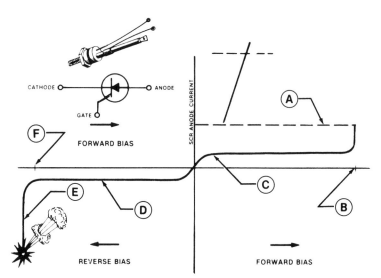

7. A(n) _____ is a three-electrode AC semiconductor switch that conducts in

both directions.

8. The _____ transistor is used primarily as a triggering device for SCRs and

triacs.

9. A(n) _____ is a bidirectional semiconductor and is used primarily as a triggering device.

10. Identify the following semiconductor devices.

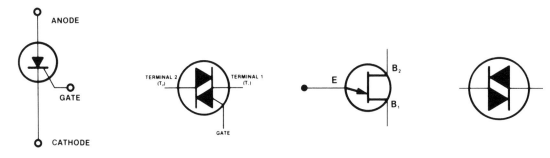

a. _____ b. _____ c. _____ d. _____

WORKSHEET 11-7

Score _____

Name _____   Class _____   Date _____

1. _____ _____ are popular because they provide a

complete circuit function in one semiconductor package.

2. The _____ amplifier is a very high gain, directly coupled amplifier that uses

external feedback to control response characteristics.

3. Identify the following gates used in digital electronics.

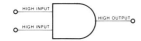

a. _____

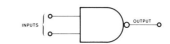

b. _____

c. _____

d. _____

4. Complete the schematic for the 555 timer and label each part.

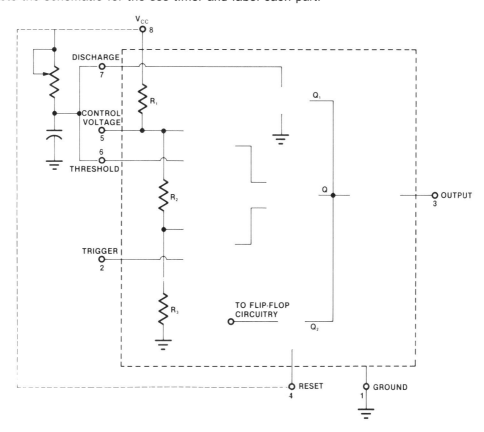

WORKSHEET 11-8

Name _____   Class _____   Date _____

1. The _____ gate is a device with an output that is high only when both of its inputs are high.

2. The _____ gate is a device with an output that is high if either or both inputs are high.

3. The _____ gate is the same as an inverted OR function.

4. The _____ gate is an inverted AND function.

5. Identify the different types of integrated circuits.

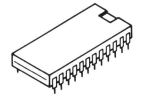 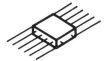 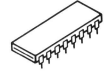

a._____ b._____ c._____ d._____ e._____

_____ 6. Laser diode

_____ 7. Pin photodiode

_____ 8. Phototransistor

_____ 9. Light activated SCR

_____10. Phototriac

_____11. Optocoupler

a. Combines effect of photodiode and transistor

b. Light sensitive gate and bidirectional

c. Provide electrical isolation between circuits

d. LASCR

e. Produces coherent light

f. Light radiation disturbs the PN junction

12. Identify the following solid state control devices.

 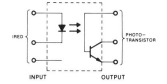

a. _____   b. _____   c. _____   d. _____

TECH-CHEK ✓ 11

Solid State Electronic Control Devices

Fill in the blanks to complete each statement.

1. _____ on a PC board are small, round conductors to which component leads are soldered.

2. _____ are used to interconnect two or more pads.

3. A PC board with multiple terminations on one end is a(n) _____ .

4. The central core of an atom is the _____ .

5. _____ is the process by which the crystal structure of an atom is altered.

6. The junction of the P-type and N-type materials in a diode is the _____ .

7. A(n) _____ is a thermally sensitive resistor.

8. A(n) _____ converts light energy into electrical energy.

9. A transducer that changes resistance with a corresponding change in pressure is called a(n) _____ .

10. _____ is the ratio of the amplitude of the output signal to the amplitude of the input signal.

11. The rating for the maximum reverse bias voltage for diodes is called the
    a. forward current
    b. reverse current
    c. peak inverse voltage
    d. depletion region

12. A(n) _____ is used to provide conduction from several sources on a PC board.
    a. edge card
    b. bus
    c. trace
    d. pad

13. Free electrons in any conductor are called _____ .
    a. neutrons
    b. carriers
    c. foils
    d. diodes

14. The SCR will remain ON as long as the current stays above a certain value. This is called
    a. avalanche current
    b. blocking current
    c. breakover voltage
    d. holding current

15. A(n) _____ provides a complete circuit function in one semiconductor package.
    a. triac
    b. diac
    c. integrated circuit
    d. breakover diac

WORKSHEET 12-1

Score _____

Name _____ Class _____ Date _____

Identify the following methods of actuating a reed relay.

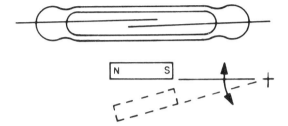

1. _____

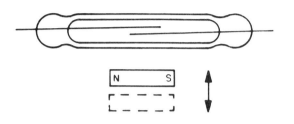

4. _____

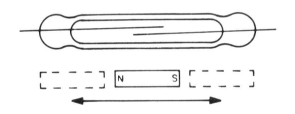

2. _____

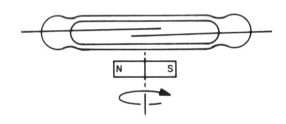

5. _____

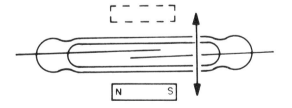

3. _____

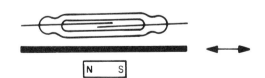

6. _____

WORKSHEET 12-2

Name _____ Class _____ Date _____

Complete the line diagrams according to the circuit information given below. Use standard lettering, numbering, and coding information. Connecting lines should be straight and the circuits neatly drawn.

Circuit 1: The control circuit must be designed so that an electromechanical relay operating at a small voltage can control a large voltage. Specifically, a 24-volt pushbutton (ON/OFF with MEMORY) should control a 240-volt solenoid.

Circuit 2: The control circuit should consist of a magnetic motor starter activated by a standard START-STOP station with MEMORY. To extend the number of contacts available, a control relay with three auxiliary contacts must be added. One NC contact is to control a green light, the other NC contact is to control a solenoid, and the NO contact is to control a red light. Overload protection should be provided for the motor.

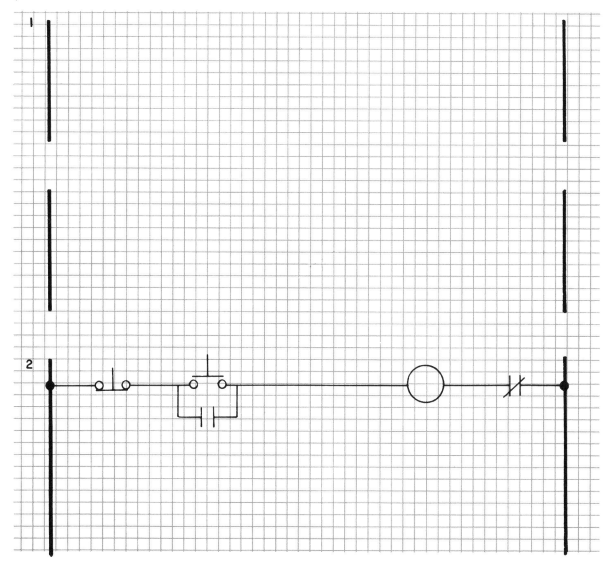

WORKSHEET 12-3

Score _____

Name _____ Class _____ Date _____

For each of the given contact arrangements shown below indicate the correct number of poles, throws, and break positions.

1._____ poles

2._____ throws

3._____ breaks

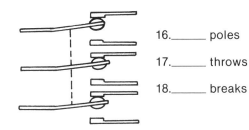

16._____ poles

17._____ throws

18._____ breaks

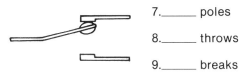

4._____ poles

5._____ throws

6._____ breaks

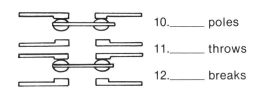

7._____ poles

8._____ throws

9._____ breaks

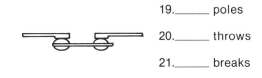

19._____ poles

20._____ throws

21._____ breaks

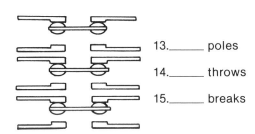

10._____ poles

11._____ throws

12._____ breaks

13._____ poles

14._____ throws

15._____ breaks

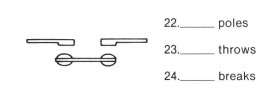

22._____ poles

23._____ throws

24._____ breaks

WORKSHEET 12-4

Name _____   Class _____   Date _____

Complete the control diagram shown below in such a way that it will contain a start pushbutton connected for MEMORY to start the motor and a stop pushbutton to stop the motor. Overload contacts should be placed so as to stop the motor in the event of an overload.

Use standard lettering, numbering, and coding information. Connecting lines should be straight and the circuit neatly drawn.

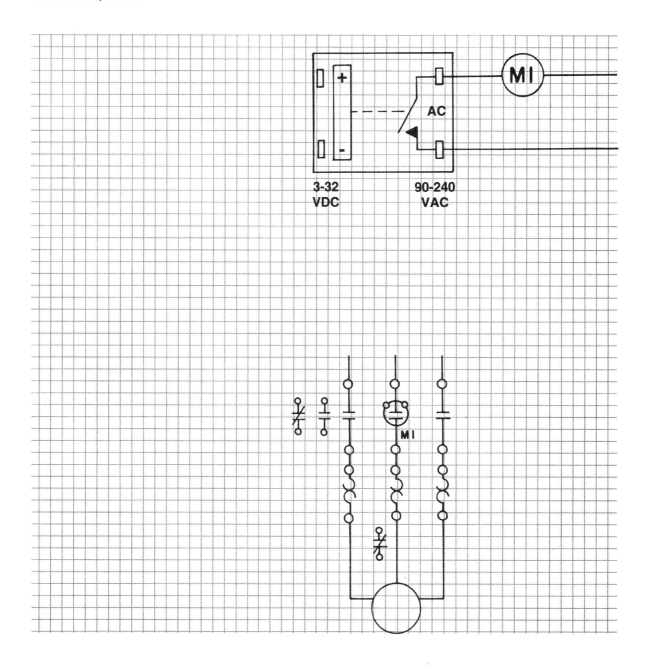

WORKSHEET 12-5

Name _____ Class _____ Date _____

Complete the control diagrams according to the circuit information given below. Use standard lettering, numbering, and coding information. Connecting lines must be straight and the circuits neatly drawn.

Circuit 1: Redraw the circuit of Worksheet 12-4, adding a timer to automatically turn off the motor after 30 minutes. The timer has a coil voltage rated at the same voltage level as the magnetic motor starter coil.

Circuit 2: Redraw the circuit of Worksheet 12-4, adding a timer to keep the motor running for 30 minutes after the stop pushbutton is pushed. The timer has a coil voltage rated at the same voltage level as the magnetic motor starter coil.

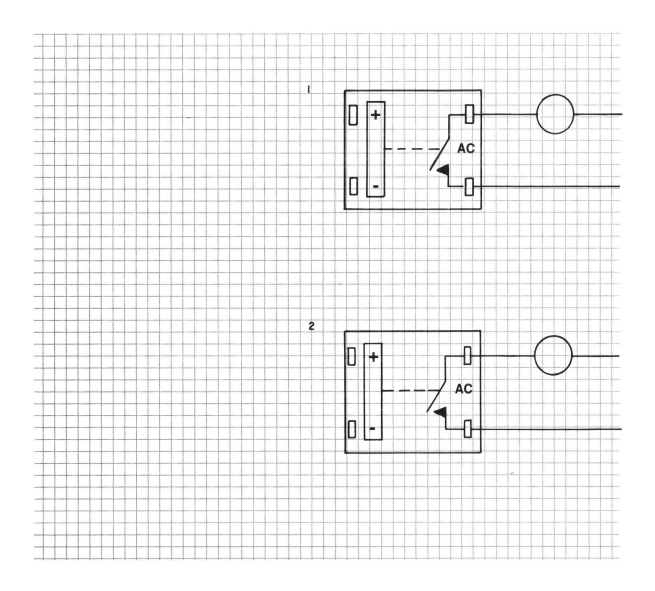

WORKSHEET 12-6

Score _____

Name _____ Class _____ Date _____

Complete the control diagram shown below, using solid state relays to design a circuit that will forward and reverse a motor. The circuit should include a FORWARD-REVERSE-STOP pushbutton and should include pushbutton interlocking.

Use standard lettering, numbering, and coding information. Connecting lines should be straight and the circuit neatly drawn.

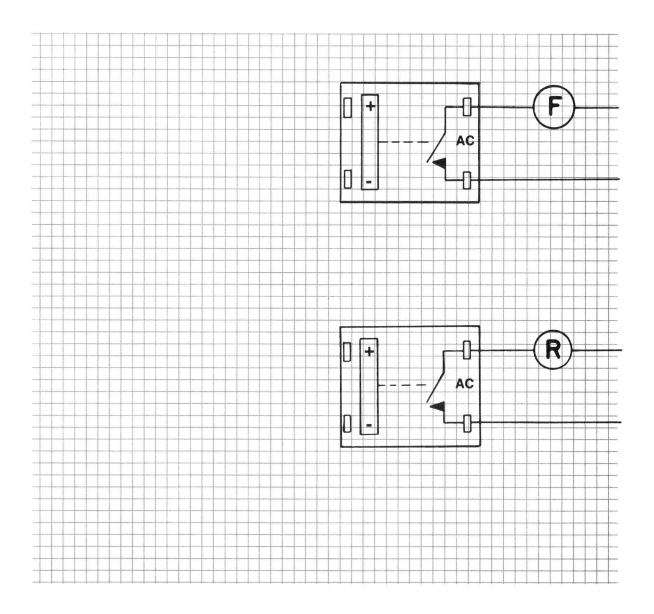

130

Solid State and Electromechanical Relays

Select the best way to complete each statement and circle a, b, c, d, e or f. Do not circle more than one choice for each statement.

1. Relays are primarily used to switch currents in the
   a. control circuit.
   b. power circuit.
   c. power distribution circuit.
   d. load circuit.

2. Relays with sets of contacts which are closed by some type of magnetic effect are called
   a. electromechanical relays.
   b. solid state relays.
   c. hybrid relays.
   d. magnetic relays.

3. Relays with no contacts, which are switched entirely by electronic devices are called
   a. electromechanical relays.
   b. solid state relays.
   c. hybrid relays.
   d. magnetic relays.

4. Relays that combine both electromechanical and solid state technology are called
   a. electromechanical ralays.
   b. solid state relays.
   c. hybrid relays.
   d. magnetic relays.

5. To the right is an illustration of
   a. perpendicular motion.
   b. parallel motion.
   c. front-to-back motion.
   d. pivoted motion.
   e. rotary motion.
   f. shielding.

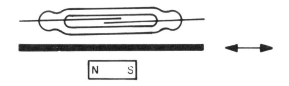

6. To the right is an illustration of
   a. perpendicular motion.
   b. parallel motion.
   c. front-to-back motion.
   d. pivoted motion.
   e. rotary motion.
   f. shielding.

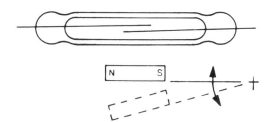

Continued

131

7. To the right is an illustration of
   a. perpendicular motion.
   b. parallel motion.
   c. front-to-back motion.
   d. pivoted motion.
   e. rotary motion.
   f. shielding.

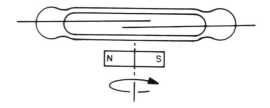

8. To the right is an illustration of
   a. perpendicular motion.
   b. parallel motion.
   c. front-to-back motion.
   d. pivoted motion.
   e. rotary motion.
   f. shielding.

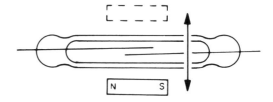

9. To the right is an illustration of
   a. perpendicular motion.
   b. parallel motion.
   c. front-to-back motion.
   d. pivoted motion.
   e. rotary motion.
   f. shielding.

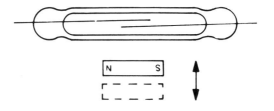

10. To the right is an illustration of
    a. perpendicular motion.
    b. parallel motion.
    c. front-to-back motion.
    d. pivoted motion.
    e. rotary motion.
    f. shielding.

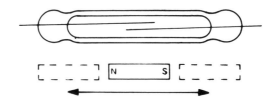

11. Below is an illustration of a
    a. direct control solid state relay.
    b. transformer isolated solid state relay.
    c. hybrid solid state relay.
    d. optically isolated solid state relay.

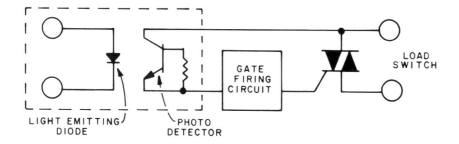

LIGHT EMITTING DIODE   PHOTO DETECTOR   GATE FIRING CIRCUIT   LOAD SWITCH

Continued

132

12. Below is an illustration of
    a. direct control solid state relay.
    b. transformer isolated solid state relay.
    c. hybrid solid state relay.
    d. optically isolated solid state relay.

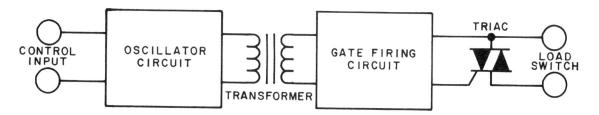

13. Below is an illustration of
    a. direct control solid state relay.
    b. transformer isolated solid state relay.
    c. hybrid solid state relay.
    d. optically isolated solid state relay.

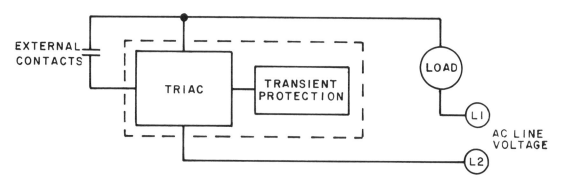

14. Below is an illustration of
    a. direct control solid state relay.
    b. transformer isolated solid state relay.
    c. hybrid solid state relay.
    d. optically isolated solid state relay.

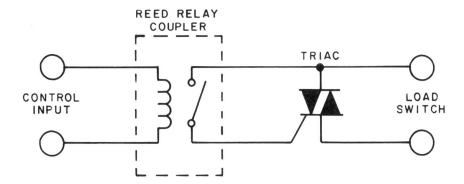

Continued

133

15. The number of completely isolated circuits that can pass through the switch at one time is the number of
   a. throws.
   b. poles.
   c. makes.
   d. breaks.

16. The number of different closed contact positions per pole that are available on a switch is the number of
   a. throws.
   b. poles.
   c. makes.
   d. breaks.

17. The number of separate contacts the switch uses to open or close each individual circuit is the number of
   a. throws.
   b. poles.
   c. makes.
   d. breaks.

18. The comparable solid state terminology for *coil voltage* as applied to electromechanical relays is
   a. holding current.
   b. holding voltage.
   c. control voltage.
   d. control current.

19. The comparable solid state terminology for *coil current* as applied to electromechanical relays is
   a. holding current.
   b. holding voltage.
   c. control voltage.
   d. control current.

20. When a capacitor and resistor are connected across the SSR output to protect against transients, the circuit is called a
   a. snubber.
   b. arc suppressor.
   c. voltage regulator.
   d. contact protector.

Fill in the blanks to complete the statements.

21. Arcless switching, zero voltage turn-on, and a high response time are all advantages of the

   _____ relay.

22. Multiple contacts, radio frequency switching, and immunity to improper functioning due to

   transients are all advantages of the _____ relay.

23. Coils, transformers, and motors are examples of _____ loads.

WORKSHEET 13-1

Score _____

Name _____ Class _____ Date _____

Referring to Data Sheet I, design a circuit that uses two light source-photoreceiver pairs to keep a hopper fill level between high and low limits. The top source-receiver pair turns a motor starter off, and the bottom source-receiver pair turns the motor starter on.

Use standard lettering, numbering, and coding information. Connecting lines should be straight and the circuit neatly drawn.

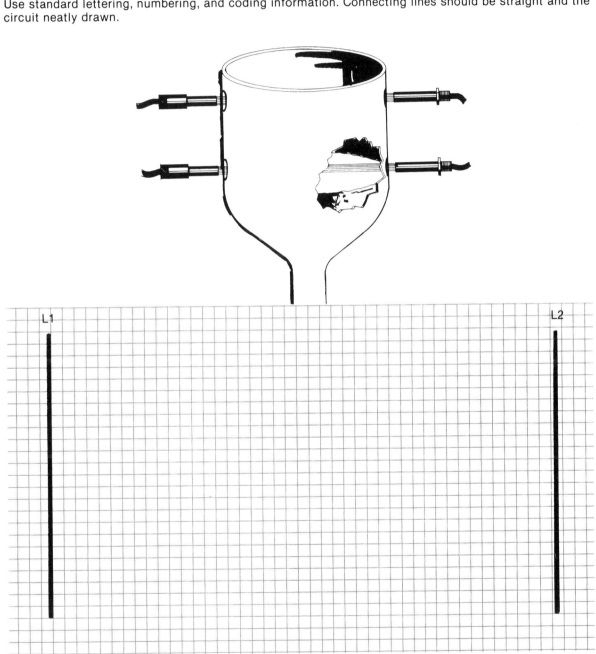

WORKSHEET 13-2

Name _____   Class _____   Date _____

Referring to Data Sheet I, design a circuit so that dark caps are checked for white liners by a photoelectric scanner. The scanner activates a solenoid valve, which controls a cylinder that rejects caps that lack liners. The solenoid is activated five seconds after the scanner sees a cap without a liner. The white liners act to stop the reflection from the tin bottom. The tin bottom acts as a reflector. You may assume there is no space between the cans.

Use standard lettering, numbering, and coding information. Connecting lines should be straight and the circuit neatly drawn.

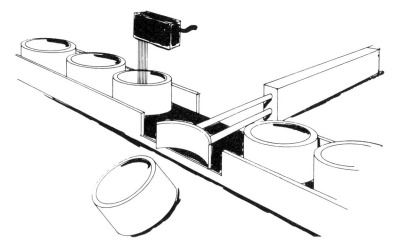

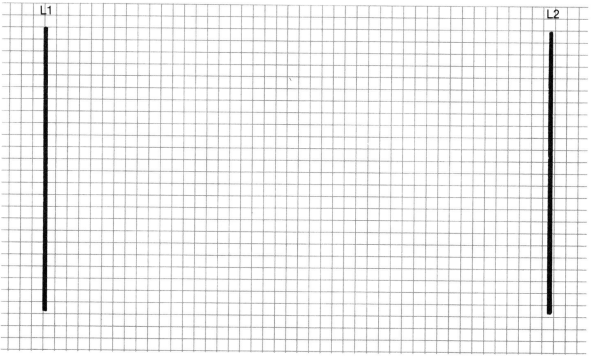

WORKSHEET 13-3

Score _____

Name _____   Class _____   Date _____

Referring to Data Sheet I, design a photoelectric control that stops a conveyor motor and fills a carton. The photoelectric control should not stop the motor until two seconds after it sees the carton. The fill process is controlled by a timer that controls a solenoid for a preset time duration. It takes 15 seconds to fill the carton. After the carton is filled, the conveyor motor is turned back on and the timer is reset. Label each type of timer as to its function (on, delay, etc.) and time setting.

Use standard lettering, numbering, and coding information. Connecting lines should be straight and the circuit neatly drawn.

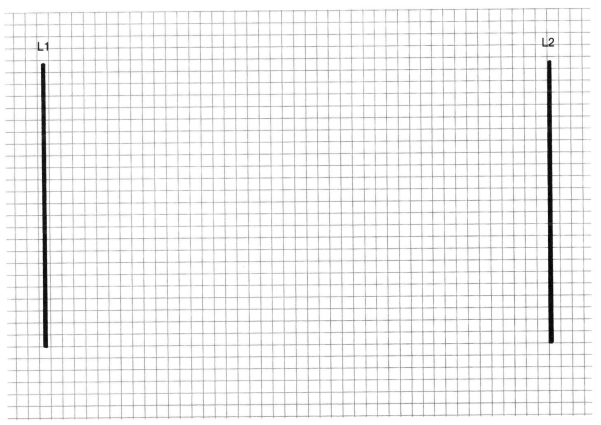

WORKSHEET 13-4

Name _____ Class _____ Date _____

Referring to Data Sheet I, design a photoelectric control circuit that turns on a glue nozzle when a carton passes by. The glue nozzle is controlled by a solenoid valve. The conveyor motor is on at all times. The glue operation should not start until two seconds after the photoelectric control sees the carton, and it should stop after five seconds, even though the carton takes nine seconds to pass the photoelectric control. This prevents glue from being sprayed on the conveyor belts. Label each type of timer as to its function (on, delay, etc.) and time setting.

Use standard lettering, numbering, and coding information. Connecting lines should be straight and the circuit neatly drawn.

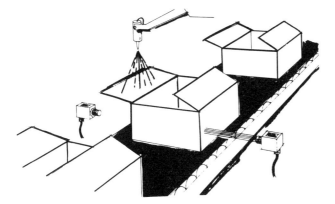

DESIRED GLUE
PATTERN ON BOX

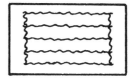

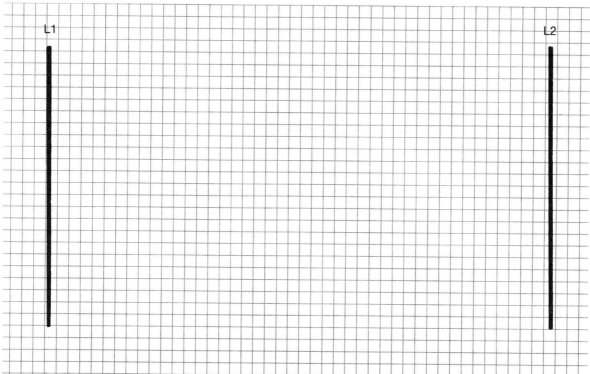

WORKSHEET 13-5

Name _____ Class _____ Date _____

Referring to Data Sheet I, design a photoelectric control circuit that sounds an alarm if there is a jam or no products moving along a conveyor system. Normally spaced products move past the photoelectric control every 1 to 3 seconds. Your circuit should sound the alarm if no product has moved past the control in 9 seconds or if a product stays in front of the control for more than 6 seconds. This will indicate a jam or that the conveyor is not moving. Label each type of timer as to its function (on, delay, etc.) and time setting.

Use standard lettering, numbering, and coding information. Connecting lines should be straight and the circuit neatly drawn.

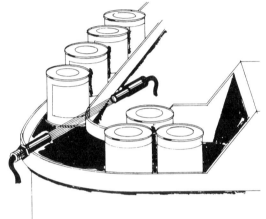

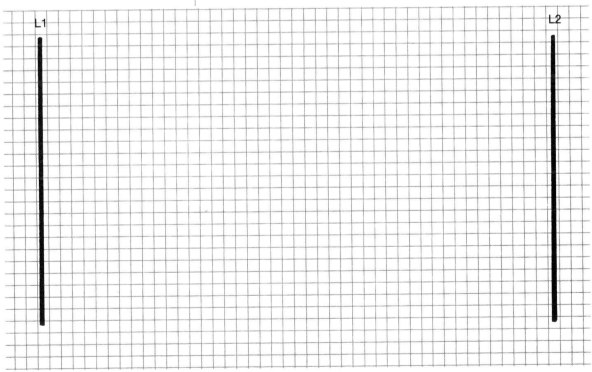

WORKSHEET 13-6

Name _____ Class _____ Date _____

Referring to Data Sheet I, design a circuit in which the position of the photoelectric control monitors the diameter of a roll of paper or fabric. The control should turn off a drive motor and sound an alarm when the roll of paper or fabric is almost empty.

Use standard lettering, numbering, and coding information. Connecting lines should be straight and the circuit neatly drawn.

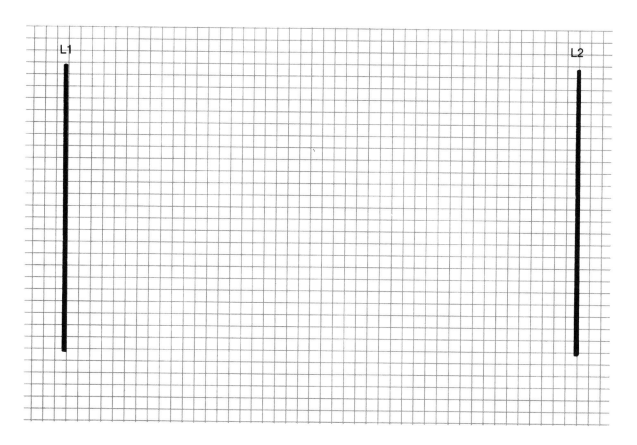

WORKSHEET 13-7

Score _____

Name _____ Class _____ Date _____

Referring to Data Sheet J, design a circuit with a proximity switch that is used to automatically stop a motor from supplying steel to a cutting press. A START pushbutton is used to start the steel in-feed, and the proximity switch or a STOP pushbutton is used to stop the steel infeed.

Use standard lettering, numbering, and coding information. Connecting lines should be straight and the circuit neatly drawn.

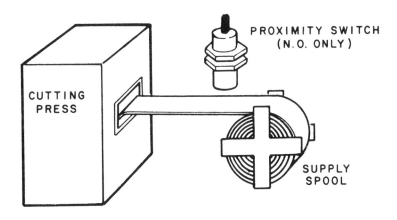

PROXIMITY SWITCH
(N.O. ONLY)

CUTTING
PRESS

SUPPLY
SPOOL

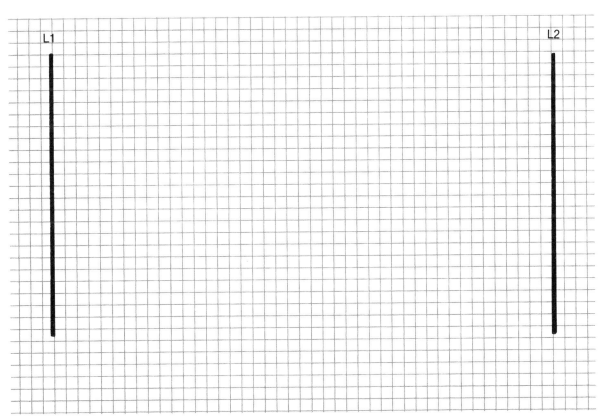

L1

L2

WORKSHEET 13-8

Name _____ Class _____ Date _____

Referring to Data Sheet I and J, design a circuit in which an alarm will sound if a bottlecap is missing. The photoelectric control is used to detect if a bottle is present and the proximity control detects if there is a cap on the bottle.

Use standard lettering, numbering, and coding information. Connecting lines should be straight and the circuit neatly drawn.

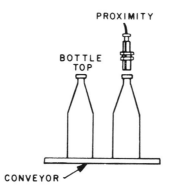

PROXIMITY

BOTTLE
TOP

CONVEYOR

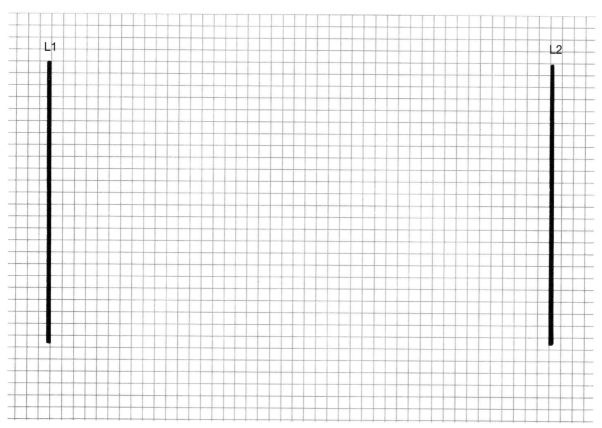

L1   L2

WORKSHEET 13-9

Name _____ Class _____ Date _____

Referring to Data Sheet J, design a circuit in which two proximity switches are used to cycle a cylinder back and forth automatically. A standard START-STOP pushbutton station with MEMORY is used to start and stop the automatic cycling. The cylinder is controlled by a double solenoid fluid power valve. One solenoid controls the cylinder out function and the other controls the cylinder in function. Power must be maintained on the solenoids to keep the cylinder moving.

Use standard lettering, numbering, and coding information. Connecting lines should be straight and the circuit neatly drawn.

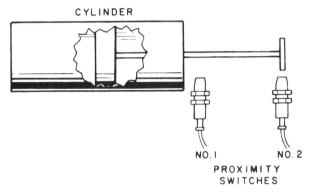

CYLINDER

NO.1   NO. 2
PROXIMITY
SWITCHES

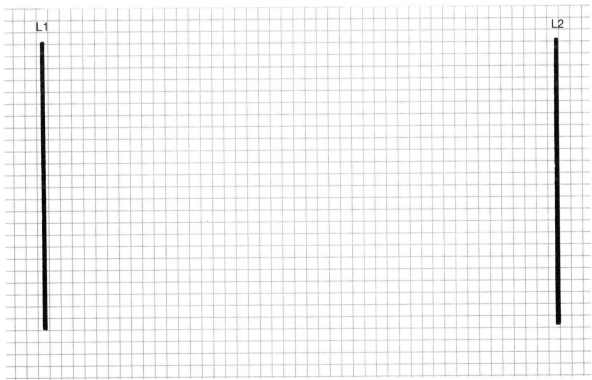

L1   L2

WORKSHEET 13-10

Name _____ Class _____ Date _____

Referring to Data Sheet K, design a circuit in which two proximity switches are used to cycle a cylinder back and forth automatically. A standard START-STOP pushbutton station with MEMORY is used to start and stop the automatic cycling. The cylinder is controlled by a double solenoid fluid power valve. One solenoid controls the cylinder out function and the other controls the cylinder in function. Power must be maintained on the solenoids to keep the cylinder moving.

Use standard lettering, numbering, and coding information. Connecting lines should be straight and the circuit neatly drawn.

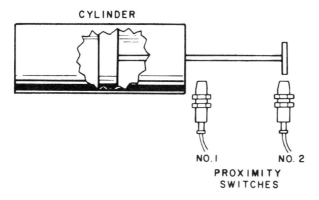

CYLINDER

NO. I   NO. 2
PROXIMITY
SWITCHES

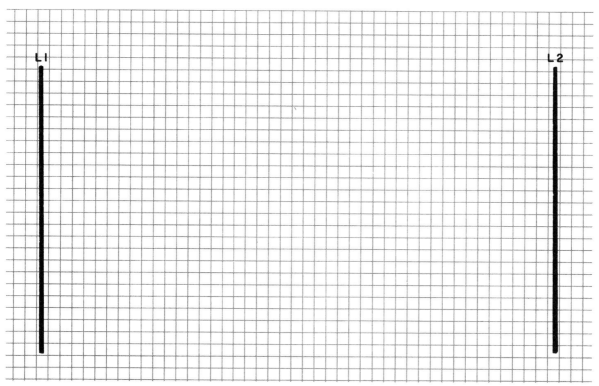

L1   L2

WORKSHEET 13-11

Name _____   Class _____   Date _____

Match the logic descriptions listed below with the logic function diagrams on the other side of the page by entering the appropriate letter in each blank space.

a. ON-OFF OPERATION: Output follows the action of the input signal.

b. ON-DELAY OPERATION: Output does not energize until input signal has been present for a preset length of time. Output de-energizes immediately when input signal is removed.

c. OFF-DELAY OPERATION: Output energizes immediately when input signal occurs, but does not de-energize until the input signal has been removed for a preset length of time.

d. ON-OFF DELAY OPERATION: Output does not energize until input signal has been present for a preset length of time and does not de-energize until the input signal has been removed for a preset length of time.

e. ONE-SHOT OPERATION: Output pulses for an adjustable time each time input signal occurs, regardless of the duration of the input signal.

f. ON-DELAY ONE-SHOT OPERATION: Output does not energize until input signal has been present for a preset length of time. Output then pulses for an adjustable time, regardless of the remaining duration of the input signal.

g. COUNTING OPERATION: Output does not energize until a preset count is reached.

h. REPEAT CYCLE TIMER OPERATION: Output continues to cycle on and off for as long as the input signal is maintained.

i. RATE SENSOR OPERATION: Output does not energize unless the input signal pulse rate falls below a preset rate. This operation can be used for underspeed detection.

j. OFF-DELAY ONE-SHOT OPERATION: Output does not energize until the input signal is removed and a preset length of time is past. Output then energizes for an adjustable time duration.

Continued

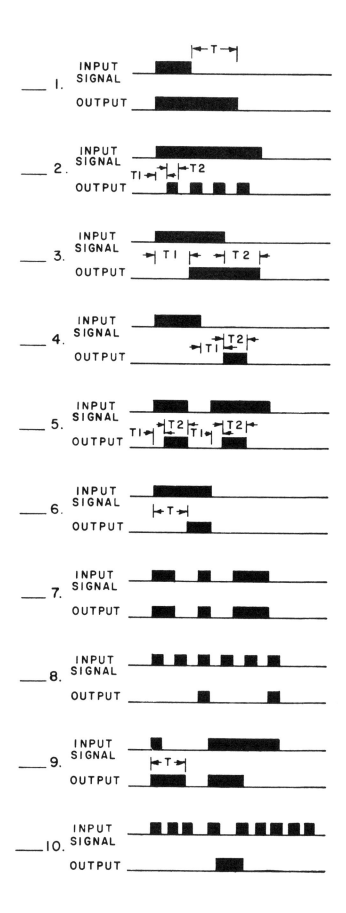

Name _____

Class _____ Date _____

Score _____

**Photoelectric and Proximity Control and
Applications**

Select the best way to complete each statement and circle a, b, c, or d.

1. When selecting a light source-photoreceiver combination, you should make sure that the combination is rated for a scanning distance
   a. exactly the size needed.
   b. smaller than the size needed.
   c. larger than the size needed.
   d. of any size.

2. As the scanning distance is increased, lamp life will
   a. increase.
   b. decrease.

3. The best type of light source to use in applications that are not easily accessible or have vibration is the
   a. incandescent type.
   b. fluorescent type.
   c. natural type.
   d. solid state type.

4. The only type of light source that should be used outdoors in photoelectric control is the
   a. modulated type.
   b. unmodulated type.
   c. filtered type.
   d. white light type.

5. If the output is energized when the light beam is not blocked and the photosensor is illuminated, the control operating mode is called
   a. light operated.
   b. dark operated.

6. If the output is energized when the light is blocked and the photosensor is not illuminated, the control operating mode is called.
   a. light operated.
   b. dark operated.

7. The operating point at which the level of light intensity will trigger the output is determined by the
   a. sensitivity adjustment.
   b. differential adustment.
   c. voltage adjustment.
   d. current adjustment.

Each of the following characteristics is an advantage of either the electromechanical (EMR) or solid state relay (SSR). Identify each by writing "EMR" or "SSR" in the blank space to the left of each characteristic.

_____ 8. Arcless switching of the load.

_____ 9. AC and DC switching.

_____10. Multipole and multithrow switching.

_____11. Zero voltage turn-on.

Continued

147

Fill in the blanks to complete each statement.

12. Below is an illustration of _____ scan.

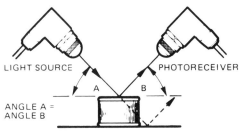

LIGHT SOURCE   PHOTORECEIVER

A   B

ANGLE A =
ANGLE B

OBJECT REFLECTS BEAM TO PHOTORECEIVER

13. Below is an illustration of _____ scan.

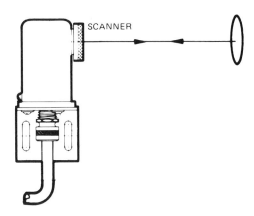

SCANNER

14. Below is an illustration of _____ scan.

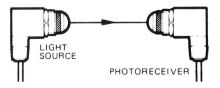

LIGHT
SOURCE

PHOTORECEIVER

15. Below is an illustration of _____ scan.

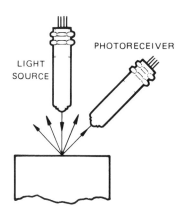

PHOTORECEIVER

LIGHT
SOURCE

WORKSHEET 14-1

Name _____   Class _____   Date _____

Following is a typical program which would be entered into the programmable controller. Using this program, draw the equivalent line diagram using standard line diagram format. Refer to Data Sheet L for information on programming in general.

Memory location	INSTRUCTION		OPERAND		Title:	date:
0	LD		IN	1		
1	OR		IN	2		
2	OR		IN	3		
3	AND	NOT	IN	4		
4	AND		IN	5		
5	=		OUT	1		
6	LD		IN	6		
7	=		OUT	2		
8	=		OUT	3		
9						

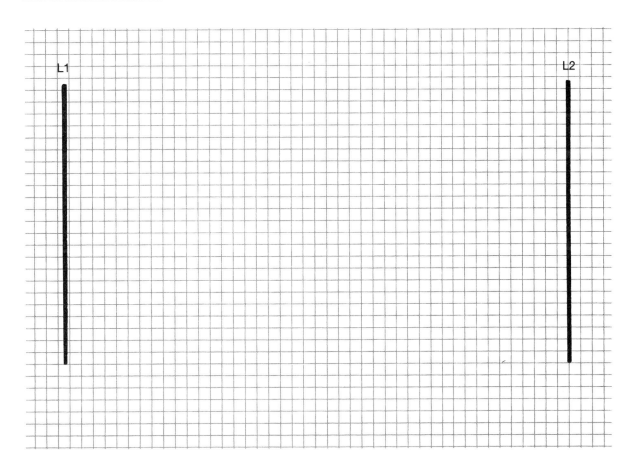

WORKSHEET 14-2

Name _____ Class _____ Date _____

Following is a typical program which would be entered into the programmable controller. Using this program, draw the equivalent line diagram using standard line diagram format. Refer to Data Sheet L for information on programming in general.

Memory location	INSTRUCTION		OPERAND		Title:	date:
0	LD		IN	1		
1	AND		IN	2		
2	=		OUT	1		
3	=		TIM	1		
4	TSET			75	7.5 SECONDS	
5	LD		TIM	1		
6	=		OUT	2		
7	=		TIM	2		
8	TSET			30	3.0 SECONDS	
9	LD		IN	3		
10	OR		TIM	2		
11	=		OUT	3		
12						

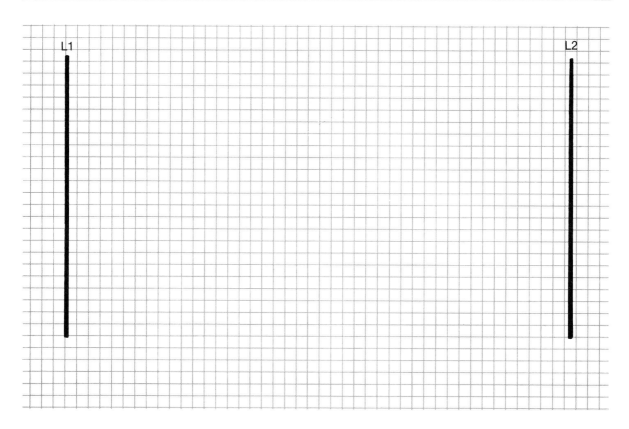

WORKSHEET 14-3

Score _____

Name _____   Class _____   Date _____

Following is a typical program which would be entered into the programmable controller. Using this program, draw the equivalent line diagram using standard line diagram format. Refer to Data Sheet L for information on programming in general.

Memory location	INSTRUCTION		OPERAND		Title: date:
0	LD		IN	1	
1	AND		IN	2	
2	OR		OUT	1	
3	AND	NOT	IN	3	
4	=		OUT	1	
5	LD		IN	4	
6	AND		IN	5	
7	OR		OUT	2	
8	AND	NOT	IN	6	
9	=		OUT	2	

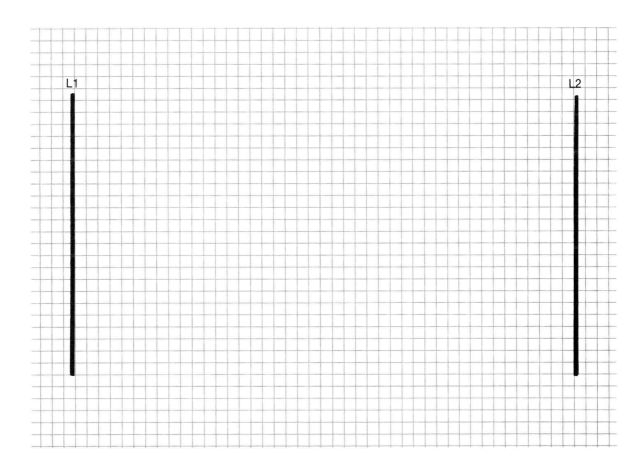

L1   L2

151

WORKSHEET 14-4

Name _____ Class _____ Date _____

Following is a typical program which would be entered into the programmable controller. Using this program, draw the equivalent line diagram using standard line diagram format. Refer to Data Sheet L for information on programming in general.

Memory location	INSTRUCTION		OPERAND		Title:	date:
0	LD		IN	1		
1	AND		IN	2		
2	AND	NOT	TIM	2		
3	=		TIM	1		
4	TSET			25	2.5 SECONDS	
5	LD		TIM	1		
6	=		OUT	1		
7	=		TIM	2		
8	TSET			50	5 SECONDS	
9	=		CNT	1		
10	LD		IN	3		
11	=		CNTR	1		
12	CSET			5		
13	LD		CNT	1		
14	=		OUT	2		

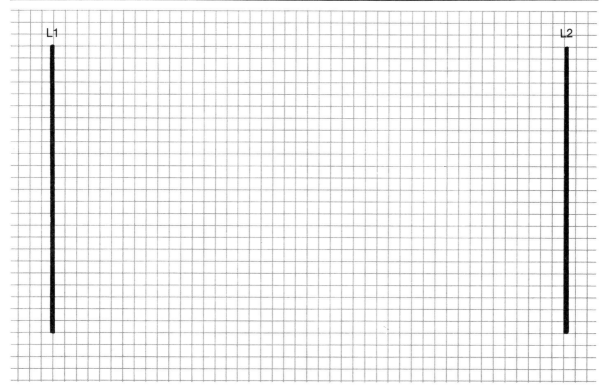

152

WORKSHEET 14-5

Score _____

Name _____ Class _____ Date _____

Illustrated below is a typical control circuit. Using this circuit, write a program that is functionally the same as this circuit. You may need to rearrange the circuit according to the information provided in Data Sheet L.

Include the following:

    A. Input 1 = PB1
    B. Input 2 = PB2
    C. Input 3 = PB3
    D. Input 4 = PB4
    E. Input 5 = Overload Contacts
    F. Output 1 = Starter Coil M1

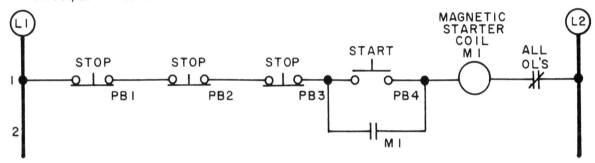

Memory location	INSTRUCTION		OPERAND		Title:	date:
0						
1						
2						
3						
4						
5						
6						
7						
8						
9						
0						
1						
2						
3						
4						
5						
6						
7						
8						
9						

WORKSHEET 14-6

Name _____   Class _____   Date _____

Illustrated below is a typical control circuit. Using this circuit, write a program that is functionally the same as this circuit. You may need to rearrange the circuit according to the information provided in Data Sheet L.

Include the following:

    A. Input 1 = PB1
    B. Input 2 = PB2
    C. Input 3 = M1 Overload Contacts
    D. Output 1 = Starter Coil M1
    D. Output 2 = Pilot Light 1

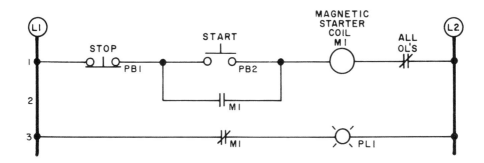

Memory location	INSTRUCTION		OPERAND		Title: date:
0					
1					
2					
3					
4					
5					
6					
7					
8					
9					
0					
1					
2					
3					
4					
5					
6					
7					
8					
9					

WORKSHEET 14-7

Score _____

Name _____ Class _____ Date _____

Illustrated below is a typical control circuit. Using this circuit, write a program that is functionally the same as this circuit. You may need to rearrange the circuit according to the information provided in Data Sheet L.

Include the following:

    A. Input 1 = PSI
    B. Input 2 = M1 Overloads
    C. Input 3 = M2 Overloads
    D. Input 4 = M3 Overloads
    E. Output 1 = M1 Starter
    F. Output 2 = M2 Starter
    G. Output 3 = M3 Starter

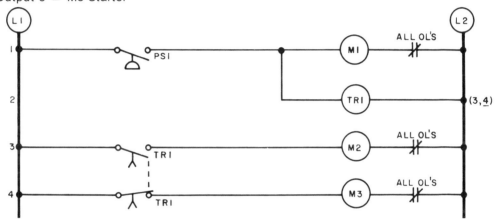

Memory location	INSTRUCTION		OPERAND		Title:	date:
0						
1						
2						
3						
4						
5						
6						
7						
8						
9						
0						
1						
2						
3						
4						
5						
6						
7						
8						
9						

WORKSHEET 14-8

Name _____   Class _____   Date _____

Illustrated below is a typical control circuit. Using this circuit, write a program that is functionally the same as this circuit. You may need to rearrange the circuit according to the information provided in Data Sheet L.

Include the following:

    A. Input 1 = Stop Pushbutton
    B. Input 2 = Start Pushbutton
    C. Input 3 = M1 Overloads
    D. Input 4 = M2 Overloads
    E. Input 5 = Flow Switch
    F. Output 1 = M1 Starter
    G. Output 2 = M2 Starter

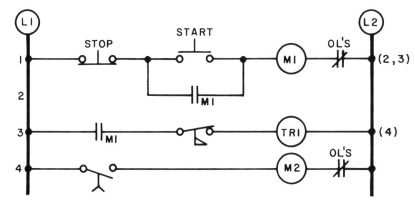

Memory location	INSTRUCTION		OPERAND		Title:	date:
0						
1						
2						
3						
4						
5						
6						
7						
8						
9						
0						
1						
2						
3						
4						
5						
6						
7						
8						
9						

TECH-CHEK ✓ 14

Name _____

Class _____ Date _____

Score _____

Programmable Controllers

1. The assembly of refrigerators is an example of _____ manufacturing.

2. The making and bottling of wine is an example of _____ manufacturing.

3. The part of the programmable controller that provides all the necessary internal voltages to the controller is the _____ section.

4. The part of the programmable controller that receives information from inputs and delivers the power to control the outputs is the _____ section.

5. The part of the programmable controller that organizes all control activity by receiving inputs, performing logical decisions according to the program, and controlling the outputs is the _____ section.

6. The part of the programmable controller that allows input information to be entered through a keyboard is the _____ section.

7. An example of a discrete input is a(n) _____ .

8. An example of a discrete output is a(n) _____ .

9. An example of a data input or output is a(n) _____ .

10. The language of programmable controller that uses a series of rungs to illustrate the logical relationship between inputs and outputs is the _____ .

11. Once a program has been developed, it can be stored outside the controller by a(n) _____ .

12. The value of a typical load register used to prevent off state leakage current turn on in a two-wire thyristor sensor is _____ ohms.

13. An example of an output load that may require surge protection is _____ .

14. The following circuit is an example of using a current _____ sensor.

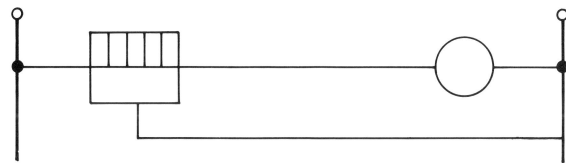

15. The following circuit is an example of using a current _____ sensor.

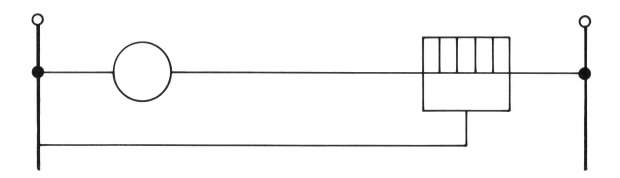

WORKSHEET 15-1

Name _____ Class _____ Date _____

Complete the wiring diagram on the other side of the page according to the line diagram shown below. When the start button is depressed, the M contactor and the time-delay relay (TR) are energized, and the motor is connected to the incoming power lines through the resistor bank. After the the time-delay relay has timed out, the timed closed contacts close and 1A contactor is energized, shorting across each of the three resistors in the resistor bank, and the motor is automatically switched to full power.

Your connecting lines should be straight and the circuit neatly drawn. Do not make any wire splices or additional terminal connections on the wiring diagram. All connections must run from terminal screw to terminal screw.

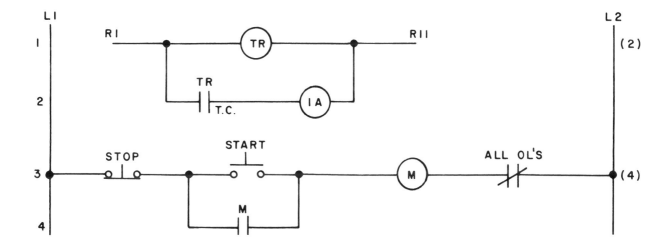

Continued

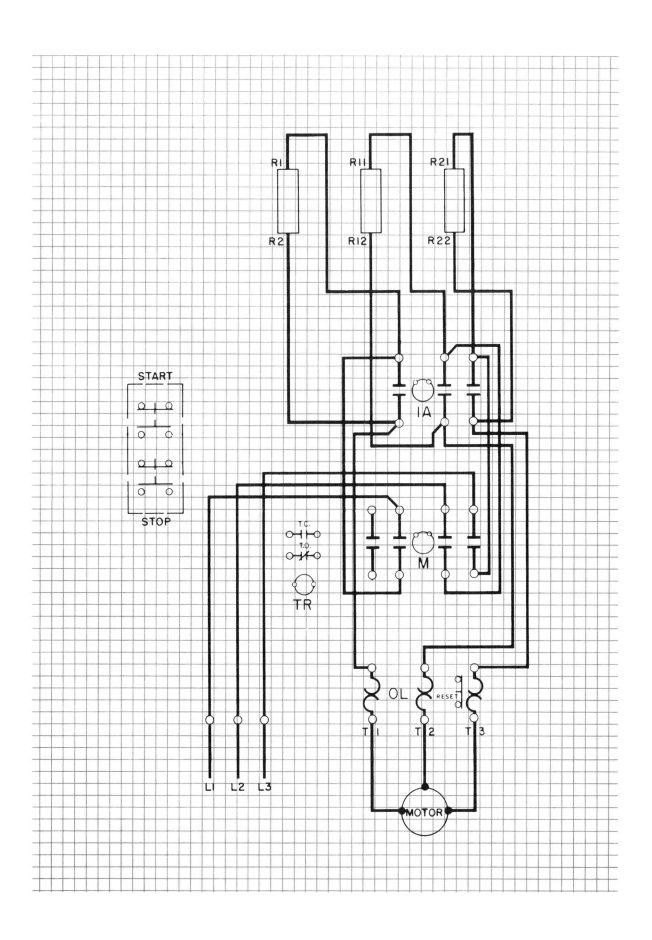

WORKSHEET 15-2

Score _____

Name _____ Class _____ Date _____

Complete the wiring diagram on the other side of the page according to the line diagram shown below. When the start button is depressed, the 1M contactor is energized first, and power is applied to motor terminals, T1, and T2, and T3. After the time-delay NO auxiliary interlock on 1M has timed out and the TC (timed closed) contacts close, the 2M contactor is energized, connecting power to the second winding, motor terminals T7, T8, and T9. The motor is stopped by depressing the stop button, which drops out both contactors. If motor terminals T4, T5, and T6 are not internally connected, they should be wired together at the terminal box as indicated by the dotted power lines at the motor.

Your connecting lines should be straight and the circuit neatly drawn. Do not make any wire splices or additional terminal connections on the wiring diagram. All connections must run from terminal screw to terminal screw.

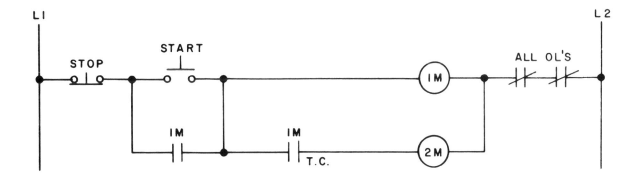

Continued

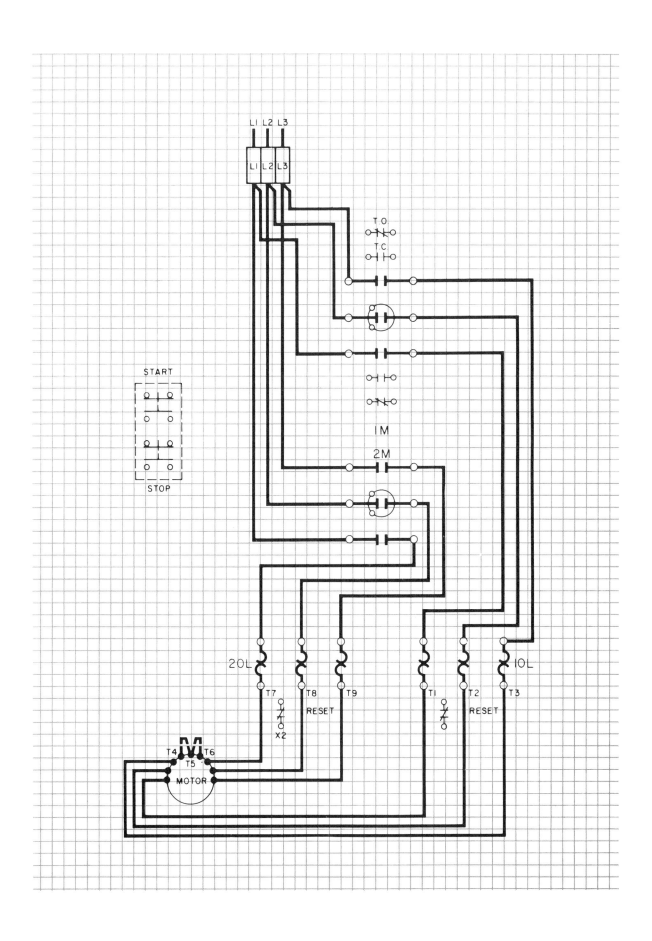

WORKSHEET 15-3

Score _____

Name _____ Class _____ Date _____

Complete the wiring diagram on the other side of the page according to the line diagram shown below. When the start button is depressed, contactors S and 1M are energized. The S contactor joins motor terminals T4, T5, and T6, and contactor 1M connects the incoming power lines to motor terminals T1, T2, and T3, causing the motor to start Wye-connected. After the time-delay NC interlock on 1M times out, the TO (timed open) contacts open, dropping out contactor S and picking up contactor 2M. The 2M contactor, upon energizing, applies power to terminals T4, T5, and T6, bringing the motor up full speed Delta-connected. The motor is stopped by depressing the stop button, which drops out all three contactors.

Your connecting lines should be straight and the circuit neatly drawn. Do not make any wire splices or additional terminal connections on the wiring diagram. All connections must run from terminal screw to terminal screw.

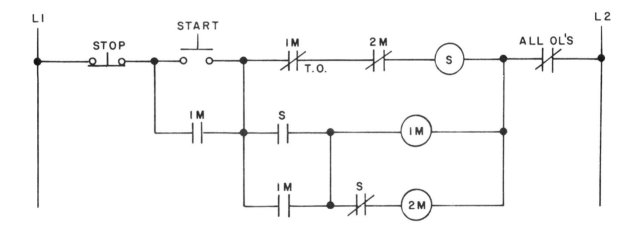

Continued

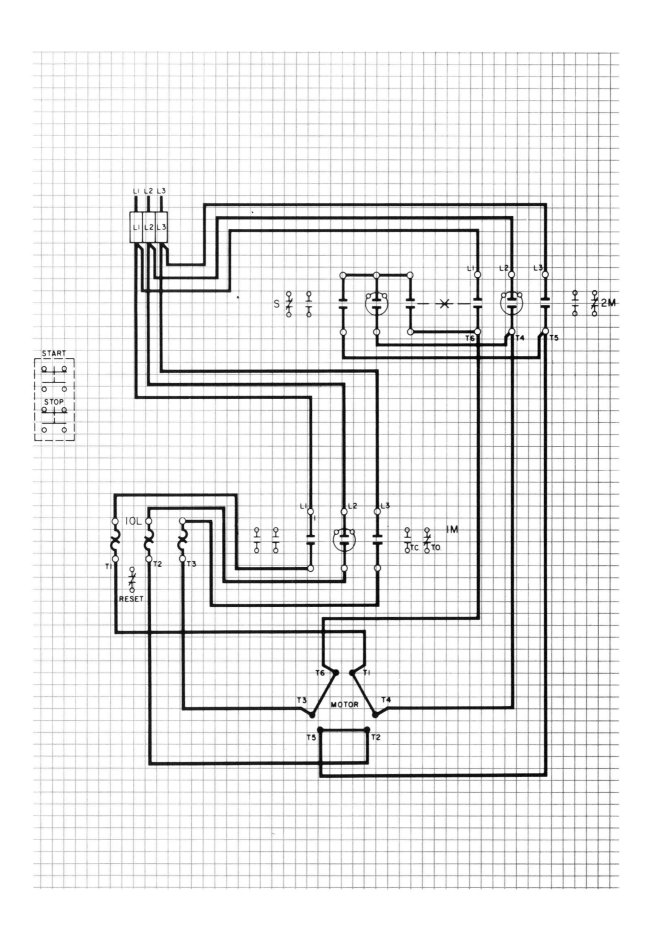

WORKSHEET 15-4

Name _____ Class _____ Date _____

Complete the wiring diagram on the other side of the page according to the line diagram shown below. When the start button is depressed, the time-delay relay (TR) and then the contactors, S and M, are energized, applying power through the windings of the autotransformer to the motor. When the time-delay relay times out, the TO (timed open) contacts open and the TC (timed closed) contacts close, the S contactor drops out, and the R contactor is energized, switching the motor to full line voltage. If the R contactor should fail to close, the sustained load on the autotransformer would cause it to overheat. The autotransformer design incorporates a thermal protector switch imbedded in the winding of each of the two transformer coils. These devices, (TPA and TPB) sense the heat rise in the coils and open their NC contacts if the temperature limits are reached. This allows full current to the lock-out relay (LR) and opens its NC contact. LR is normally shorted out by the thermal protector switches. The lock-out relay has to be hand-reset to restore power to the line. This is a safety feature to protect the intermittent duty windings of the autotransformer from overheating.

Your connecting lines should be straight and the circuit neatly drawn. Do not make any wire splices or additional terminal connections on the wiring diagram. All connections must run from terminal screw to terminal screw.

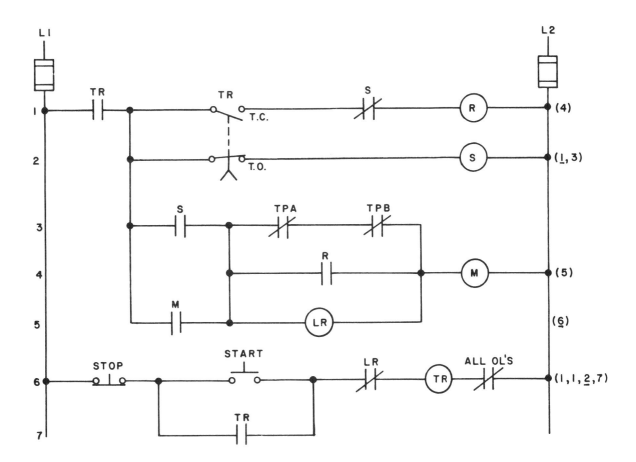

Continued

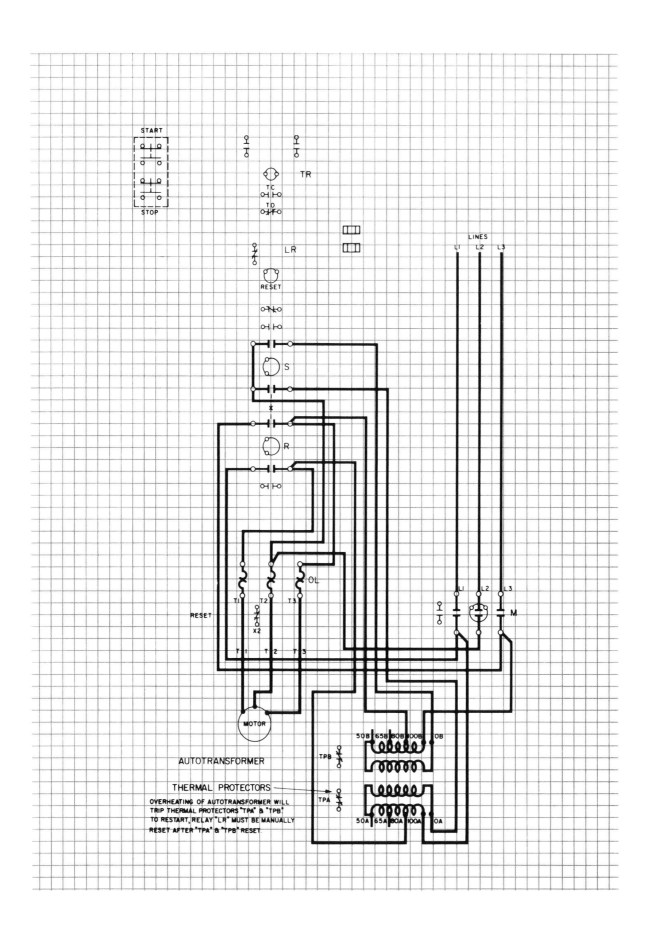

START

STOP

TR

TC

TD

LINES

L1 L2 L3

LR

RESET

S

R

OL

T1 T2 T3

RESET

X2

L1 L2 L3

M

T 1 T 2 T 3

MOTOR

AUTOTRANSFORMER

50B 65B 80B 100B 0B

TPB

THERMAL PROTECTORS

OVERHEATING OF AUTOTRANSFORMER WILL
TRIP THERMAL PROTECTORS "TPA" & "TPB"
TO RESTART, RELAY "LR" MUST BE MANUALLY
RESET AFTER "TPA" & "TPB" RESET.

TPA

50A 65A 80A 100A 0A

WORKSHEET 15-5

Name _____ Class _____ Date _____

Complete the line diagrams according to the circuit information given below. Use standard lettering, numbering, and coding information. Connecting lines should be straight and the circuits neatly drawn.

Circuit 1: In the line diagram of Worksheet 15-2, magnetic motor starter 1M had a time-delay auxiliary contact installed along with the instantaneous contacts. Since not all motor starters have a time-delay auxiliary contact add-on option, or one may not be available at the time of installation, a separate timer is often used. Redraw the line diagram of Worksheet 15-2, showing how a separate timer could be added to the circuit when a time-delay auxiliary contact is not used.

Circuit 2: Since reduced voltage starting is used to control large motor types, it is often required that the motor not be restarted for a given amount of time after it has turned off. Redraw the line diagram of Circuit 1, adding a second timer that does not allow the motor to be restarted until it has stopped for 10 minutes. Since this type of operation is usually automatic, as in the case of air-conditioning, replace the pushbutton station with a temperature controller.

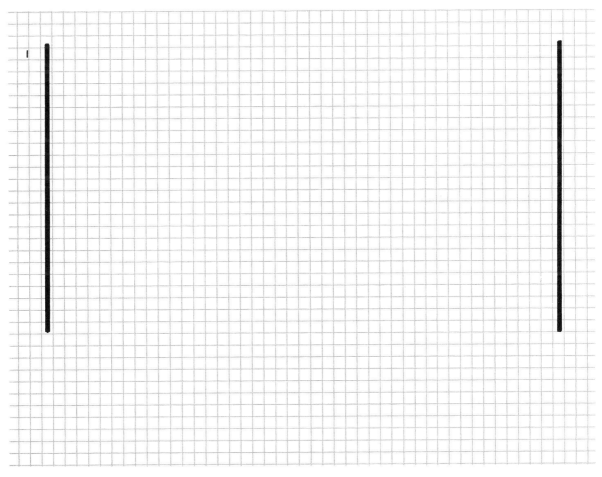

Continued

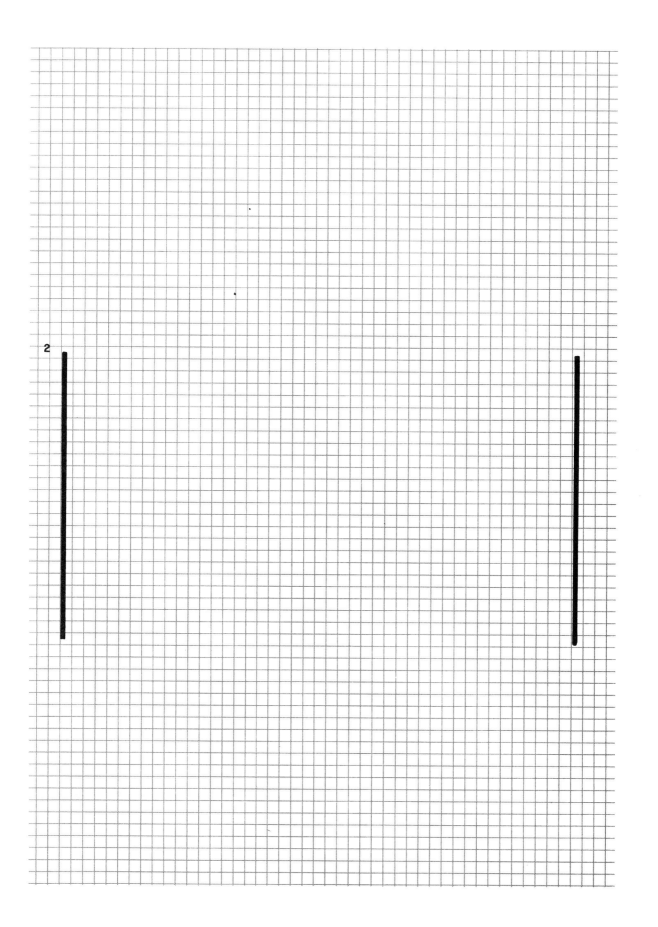

2

WORKSHEET 15-6

Score _____

Name _____ Class _____ Date _____

Complete the line diagram below by redrawing the line diagram of Circuit 2 on Worksheet 15-5, adding a cooling fan motor to run for 10 minutes after the reduced voltage starting motor has turned off.

Use standard lettering, numbering, and coding information. Connecting lines should be straight and the circuit neatly drawn.

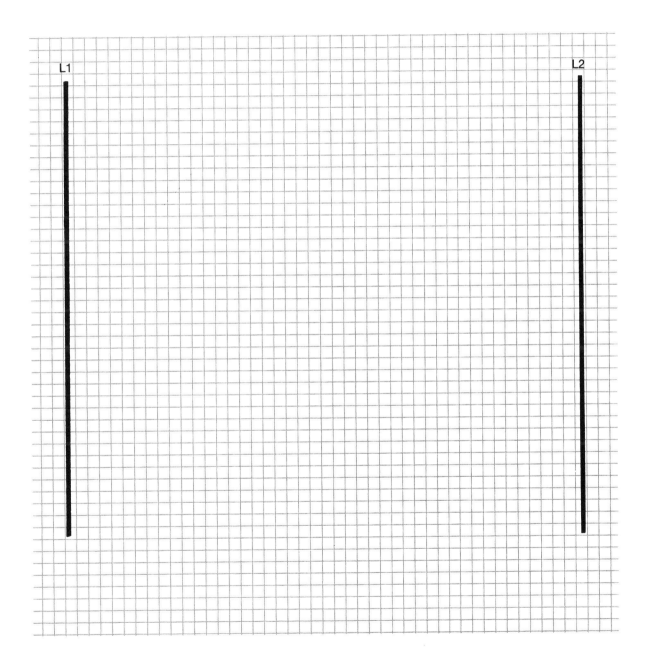

WORKSHEET 15-7

Name _____ Class _____ Date _____

Illustrate in a line diagram how a load guard power factor monitor can be added to a standard start/stop pushbutton circuit. The start/stop station is controlling the grinder illustrated below and the load guard will automatically stop the motor upon a jam or overload. Refer to Data Sheet M for the connection and use of the load guard monitor.

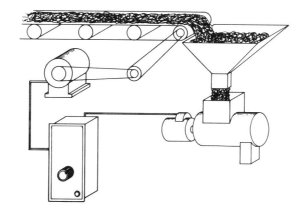

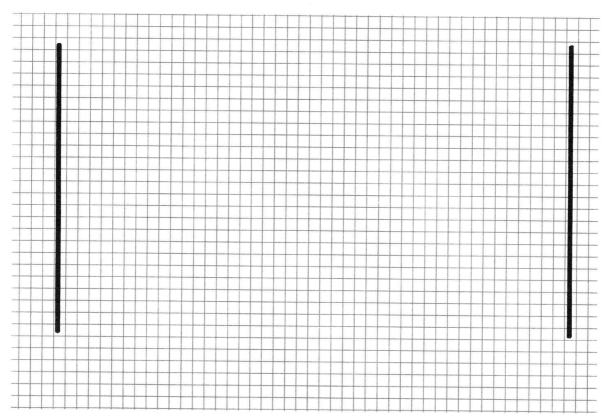

AC Reduced Voltage Starters

Select the best way to complete each statement and circle a, b, c, or d.

1. Reduced voltage starting is used as a means of
   a. reducing the starting current.
   b. speed control.
   c. full voltage starting.
   d. starting difficult loads.

2. The part of a DC motor that is used to reverse the direction of current flow in the armature coils is the
   a. brush set.
   b. power supply.
   c. commutator.
   d. armature poles.

3. The part of a DC motor that is used to provide the contact to the external power supply is the
   a. brush set.
   b. armature.
   c. commutator.
   d. field poles.

4. Sparking at the brushes of larger DC motors is reduced by
   a. commutators.
   b. additional main poles.
   c. interpoles.
   d. insulator.

5. The DC motor, unlike the AC motor, may need reduced voltage starting in order to
   a. reduce the current.
   b. reduce the torque.
   c. protect the electrical environment.
   d. protect the motor.

6. The speed of an AC squirrel cage motor can be changed by
   a. changing the number of poles.
   b. changing the supply voltage.
   c. changing the supply current.
   d. adding interpoles connected to the armature.

Fill in the blanks to complete each statement.

7. When voltage is reduced to a motor, so are the current and _____

   of the motor

8-9. In an AC squirrel-cage motor, the fixed frame is called the (8)_____ and

   the rotating part is called the (9)_____.

Continued

171

10. The current required by the motor to produce full load torque at the motor's rated speed is called _____.

11. The steady-state current taken from the power lines when the motor is started is called

_____.

12-16. The five methods used in reduced voltage starting are (12)_____,

(13)_____, (14)_____,

(15)_____ and (16)_____.

17-18. Two types of reduced voltage starting that are not adjustable to more than two steps without additional circuitry are the (17)_____ and the

(18)_____.

19. The reduced voltage starting method which is adjustable through its entire range is the

_____.

WORKSHEET 16-1

Name _____ Class _____ Date _____

Complete the wiring diagram on the other side of the page according to the line diagram shown below. The motor is started in normal fashion by pushing the start button. As it accelerates, centrifugal force closes the NO contacts of the plugging switch, picking up relay 1CR, which holds in through its own NO contacts. When the stop button is depressed, the forward contactor (F) drops out and picks up the reverse contactor (R) through its NC auxiliary contacts and through the contacts of relay 1CR. The reverse contactor applies reverse power to the motor until its speed is sufficiently reduced to allow the contacts of the plugging switch to open. The opening of these contacts causes both 1CR and R to drop out completely, disconnecting the motor from the line. Relay 1CR is used in this circuit as a safety interlock to prevent the starter from energizing when the motor shaft is turned by hand.

Your connecting lines should be straight and the circuit neatly drawn. Do not make any wire splices or additional terminal connections on the wiring diagram. All connections must run from terminal screw to terminal screw.

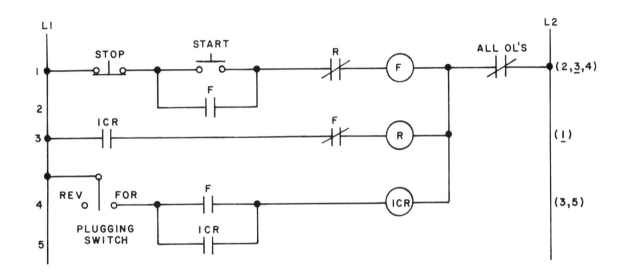

Continued

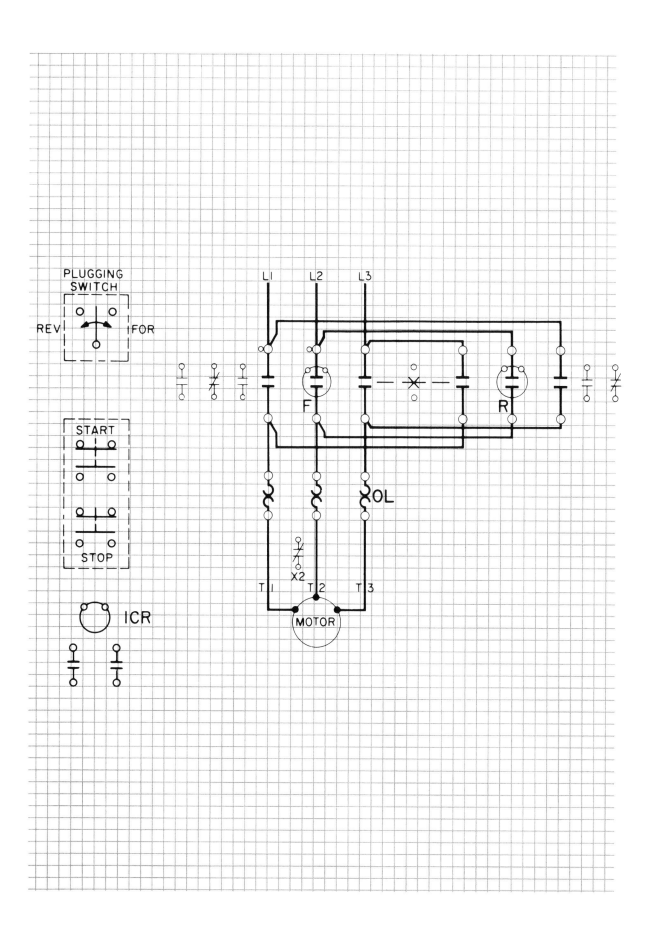

WORKSHEET 16-2

Score _____

Name _____ Class _____ Date _____

Complete the wiring diagram on the other side of the page according to the line diagram shown below. The circuit provides for a compelling, two-speed starter (fast and slow speed) with two thermal overload relays, one for each speed. This controller is internally wired to compel the operator to start the motor at the slow speed. It cannot be switched to the fast speed until after the motor is running. When the slow button is depressed, the slow speed contactor and the control relay (FR) are energized. If the fast button is depressed in an attempt to start the motor, nothing will happen as the NC contacts of the control relay will prevent the high-speed contactor from energizing. Once the motor is running, depressing the fast button will automatically drop out the slow speed contactor and pick up the high speed through the NC contacts of the slow speed contactor and the NO contacts of the control relay. With the control wired as indicated, the starter cannot be switched from fast to slow without depressing the stop button first.

Your connecting lines should be straight and the circuit neatly drawn. Do not make any wire splices or additional terminal connections on the wiring diagram. All connections must run from terminal screw to terminal screw.

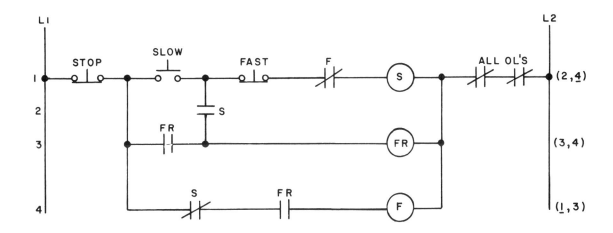

Continued

175

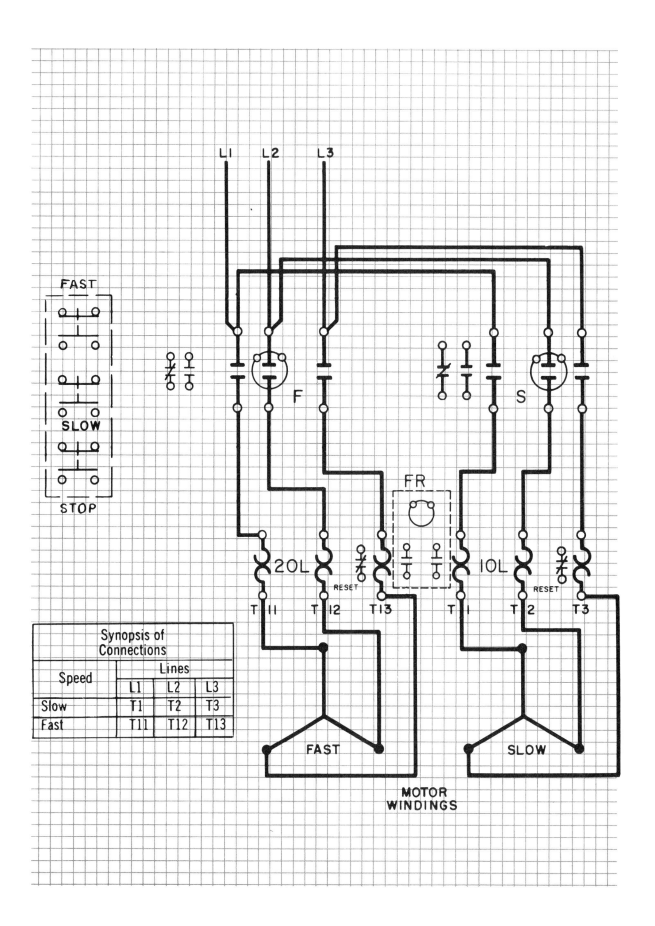

FAST

SLOW

STOP

L1 L2 L3

F

S

FR

20L

RESET

10L

RESET

T11 T12 T13

T1 T2 T3

Synopsis of Connections			
Speed	Lines		
	L1	L2	L3
Slow	T1	T2	T3
Fast	T11	T12	T13

FAST

SLOW

MOTOR
WINDINGS

WORKSHEET 16-3

Name _____   Class _____   Date _____

Complete the wiring diagram on the other side of the page according to the line diagram shown below.

The circuit provides for a multi-speed starter connected for operation with a reconnectable, constant horsepower motor. The control is a three-element FAST-SLOW-STOP pushbutton station connected for starting in either the fast or slow speed. To change the speed from fast to slow, you must first push the stop button.

Your connecting lines should be straight and the circuit neatly drawn. Do not make any wire splices or additional terminal connections on the wiring diagram. All connections must run from terminal screw to terminal screw.

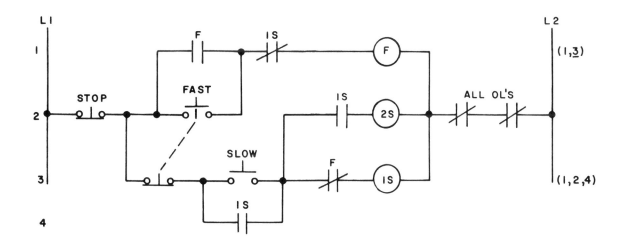

Continued

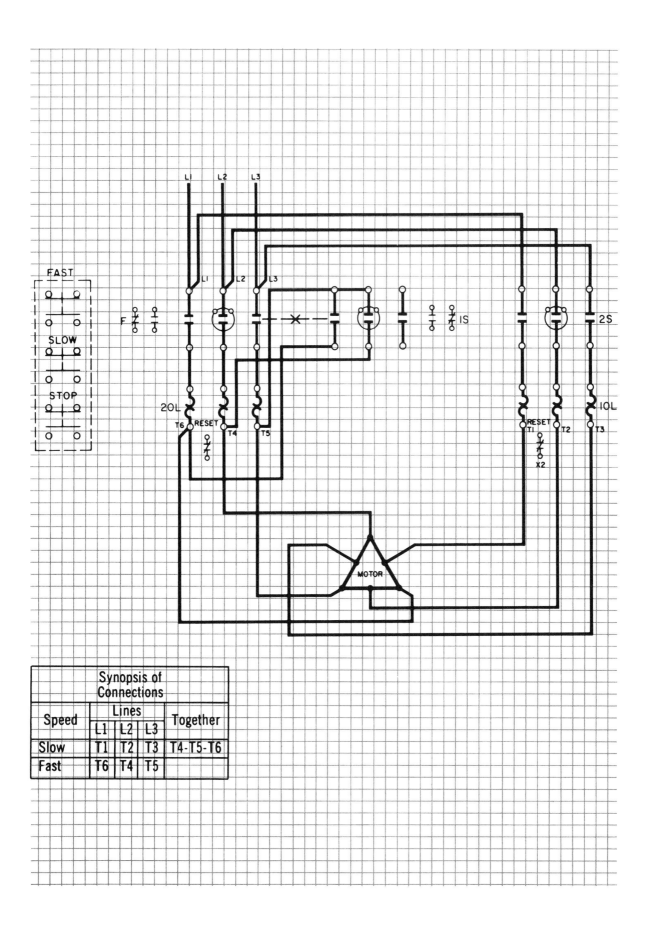

Synopsis of Connections				
Speed	Lines			Together
	L1	L2	L3	
Slow	T1	T2	T3	T4-T5-T6
Fast	T6	T4	T5	

WORKSHEET 16-4

Score _____

Name _____ Class _____ Date _____

Complete the wiring diagram on the other side of the page according to the line diagram shown below.

The circuit provides for a selective, multi-speed starter with two thermal overload relays, one for each speed. The control illustrated is a standard three-element FAST-SLOW-STOP pushbutton station. When connected as shown in the wiring diagram and line diagram, the motor can be started at either the fast or slow speed but it cannot be switched from fast speed to slow without first pressing the stop button.

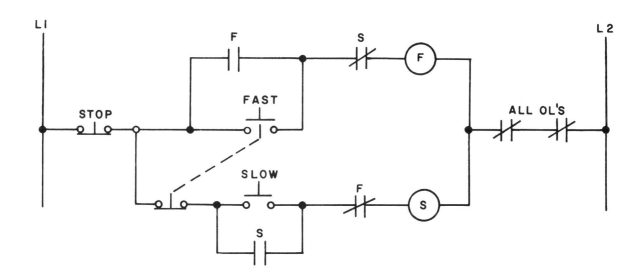

Continued

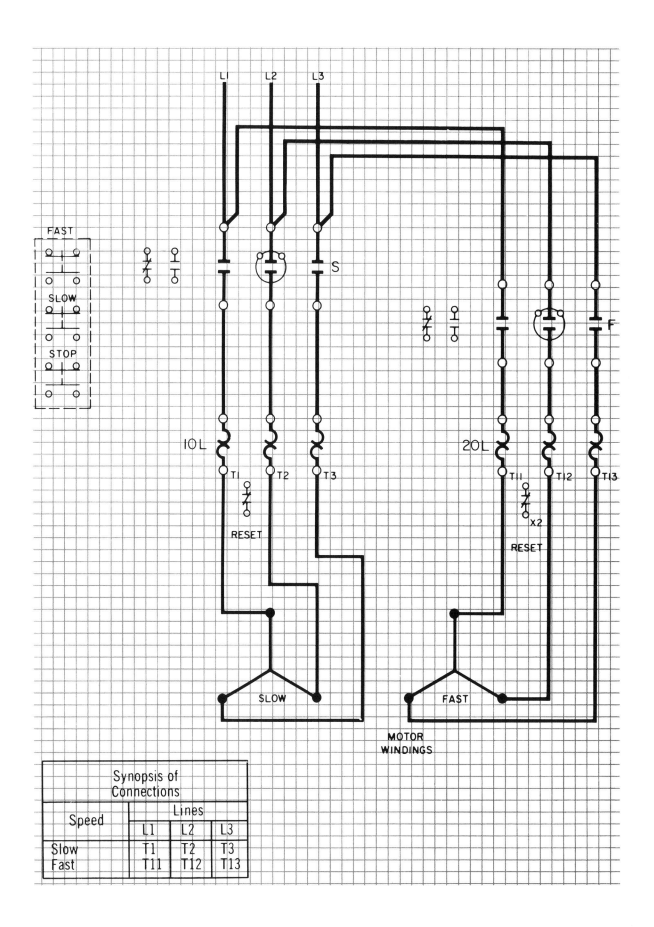

FAST

SLOW

STOP

L1 L2 L3

S

F

10L

20L

T1 T2 T3

T11 T12 T13

x2

RESET

RESET

SLOW

FAST

MOTOR
WINDINGS

Synopsis of Connections			
Speed	Lines		
	L1	L2	L3
Slow	T1	T2	T3
Fast	T11	T12	T13

WORKSHEET 16-5

Name _____   Class _____   Date _____

Complete the wiring diagram on the other side of the page according to the line diagram shown below. The control is a standard three-element FAST-SLOW-STOP pushbutton station connected for starting at either the fast or slow speed. It cannot be switched from fast speed to slow without first pressing the stop button.

Your connecting lines should be straight and the circuit neatly drawn. Do not make any wire splices or additional terminal connections on the wiring diagram. All connections must run from terminal screw to terminal screw.

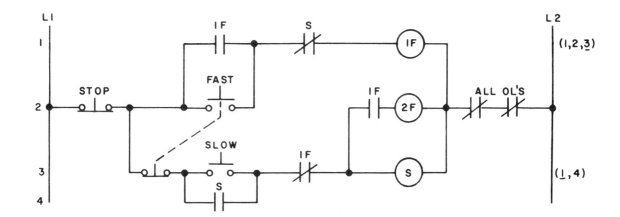

Continued

181

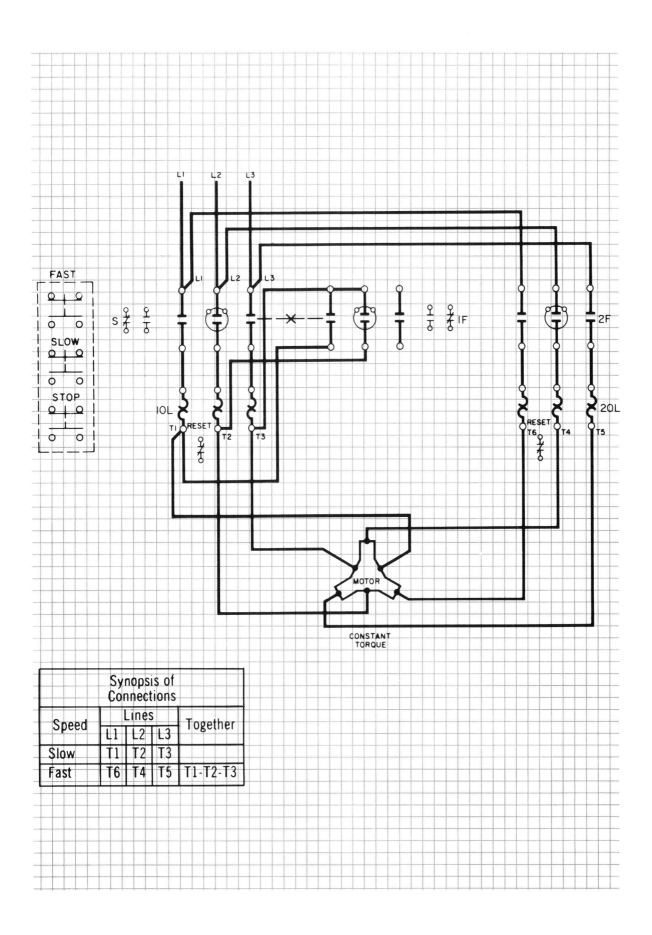

	Synopsis of Connections			
	Lines			Together
Speed	L1	L2	L3	
Slow	T1	T2	T3	
Fast	T6	T4	T5	T1-T2-T3

WORKSHEET 16-6

Name _____ Class _____ Date _____

Design a line diagram for compelling circuit logic, using three motor starters (low, medium, and high). The motor must be started at low speed before changing to medium speed, and it must be running at medium speed before changing to high speed. Two control relays should be used to complete this circuit. Overload protection should be provided at all speeds.

Use standard lettering, numbering, and coding information. Connecting lines should be straight and the circuit neatly drawn.

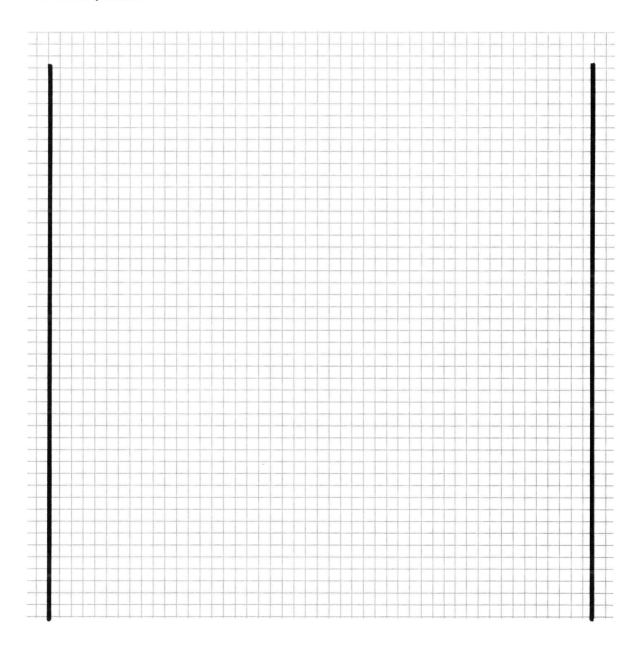

WORKSHEET 16-7

Score _____

Name _____ Class _____ Date _____

Design a circuit in which a friction brake can be applied to a DC shunt motor wired for dynamic braking. The friction brake is to be applied three seconds after the motor is turned off, allowing the dynamic braking action to slow the motor first. The friction brake is to remain on, holding the load, until the motor is started again.

Use standard lettering, numbering, and coding information. Connecting lines should be straight and the circuit neatly drawn.

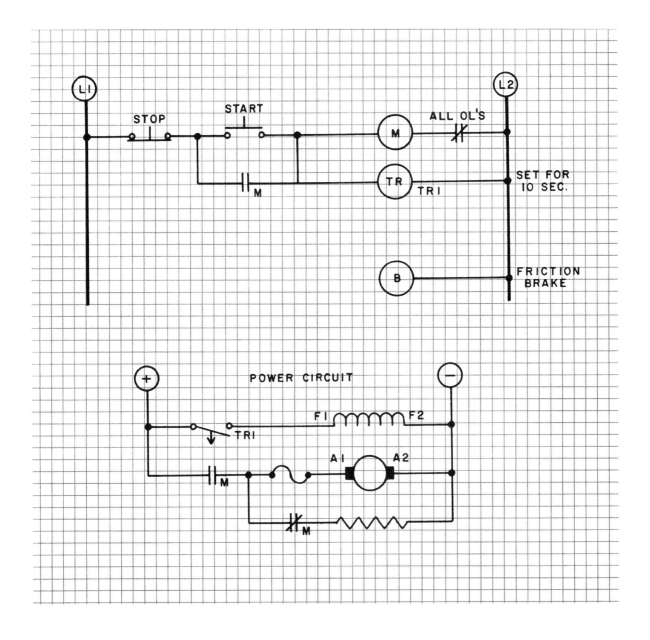

WORKSHEET 16-8

Name _____ Class _____ Date _____

Design a circuit in which a temperature switch turns on and off a heating element and circulating fan based on a given temperature. Build into this circuit a plugging switch that would automatically turn off the heating element contactor if the circulating fan were no longer operating. Overload protection should be provided for the motor.

Use standard lettering, numbering, and coding information. Connecting lines should be straight and the circuit neatly drawn.

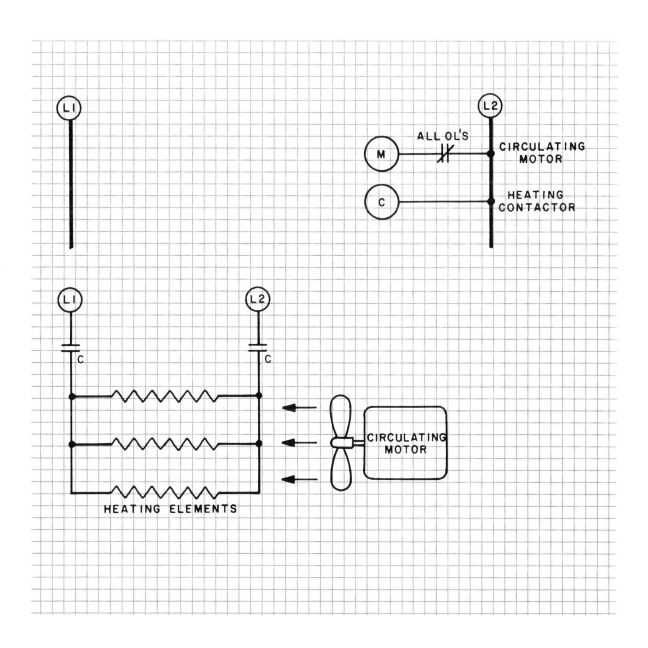

WORKSHEET 16-9

Name _____ Class _____ Date _____

Answer the following questions, showing all of your work in solving each problem at the bottom of the worksheet or on a separate sheet of paper.

    1. A drive motor runs a 3600 RPM with a 12-inch pulley, and the driven machine runs at 1800 RPM. What size would the driven pulley be?

    2. A drive motor runs at 3600 RPM with a 10-inch pulley. The driven machine has a 30-inch pulley. At what speed would the driven machine run?

    3. A drive motor runs at 900 RPM. The driven machine runs at 360 RPM with a 10-inch pulley. What size would the drive pulley be?

    4. A driven machine runs at 640 RPM with a 15-inch pulley. The drive motor has an 8-inch pulley. At what speed would the drive motor run?

ANSWERS

SPACE FOR STUDENT CALCULATIONS

1. _____

2. _____

3. _____

4. _____

TECH-CHEK ✔ 16

Name _____

Class _____ Date _____

Accelerating and Decelerating Methods and Circuits

Score _____

Select the best way to complete each statement and circle a, b, c, or d.

1. Braking torque developed with friction brakes is directly proportional to the
   a. speed of the motor.
   b. surface area and spring pressure.
   c. applied voltage.
   d. type of solenoid used.

2. The advantage of using friction brakes is
   a. low maintenance.
   b. simplified maintenance.
   c. their ability to control loads that are stopped often.
   d. their ability to be connected to any voltage.

3. In electric braking the amount of braking force is varied by changing the
   a. spring pressure.
   b. time the motor is connected in reverse.
   c. surface area of brake.
   d. applied voltage.

4. Dynamic braking is usually applied to DC motors because
   a. DC motors reverse faster than AC motors.
   b. there must be access to the rotor.
   c. AC motors do not have stationary fields.
   d. AC power cannot develop a retarding torque.

5. Friction brakes are sometimes used with dynamic braking because
   a. dynamic braking cannot hold a stopped load.
   b. friction brakes are less expensive.
   c. dynamic braking uses too much power.
   d. AC brakes cannot be used with DC.

6. An example of a load that would require a constant torque/variable horsepower motor is
   a. paper roll machine.
   b. clock.
   c. fan.
   d. conveyor.

7. An example of a load that would require a constant horsepower/variable torque motor is
   a. paper roll machine.
   b. clock.
   c. fan.
   d. conveyor.

8. An example of a load that would require a variable torque/variable horsepower motor is a
   a. paper roll machine.
   b. clock.
   c. fan.
   d. conveyor.

Continued

Fill in the blanks to complete each statement.

9. For a timing relay to plug a motor to a stop, a _____ delay timer is used.

10. A motor used for plugging should have a service factor of _____ or more.

11. Electric braking is a method of braking in which a _____ current is applied to the stationary windings of a motor after the voltage is removed.

12. In dynamic braking, the smaller the resistance of the resistor used, the _____ the rate of energy dissipation.

13. If a motor is required to lift a 100-pound load 200 feet in one minute, a _____ size motor is required.

14. If a motor is required to lift a 100-pound load 500 feet in one minute, a _____ size motor is required.

15. In a DC motor, if the voltage to the armature is reduced, the motor speed is _____.

16. In a DC motor, if the voltage to the field is reduced, the motor speed is _____.

17-18. To change the speed of an AC induction motor, the (17)_____ or (18)_____ must be changed.

19. An AC induction motor with four poles will run at a synchronous speed of _____ RPM.

20. A circuit that starts the motor in low speed and automatically brings the motor to high speed after the HIGH pushbutton is pressed is called _____ circuit logic.

21. A circuit that requires the motor to start at low speed before the motor can be put into high speed is called _____ circuit logic.

22. A _____ changes the standard 60 hertz AC power into almost any desired frequency.

23. A _____ changes a DC voltage into AC variable frequency.

24. If a drive motor runs at 1800 RPM with a 10-inch pulley, and the driven machine is to run at 1000 RPM, the driven pulley size would be _____ inches.

WORKSHEET 17-1

Score _____

Name _____   Class _____   Date _____

Referring to Data Sheet N, complete the wiring diagram and line diagram below, showing how this monitor (SY 185) can be used to protect against phase loss or phase angle error (adjustable on this model from 5 to 15 degrees). Phase angle error will take place in an unbalanced system. The monitor is rated for the same voltage as the power lines.

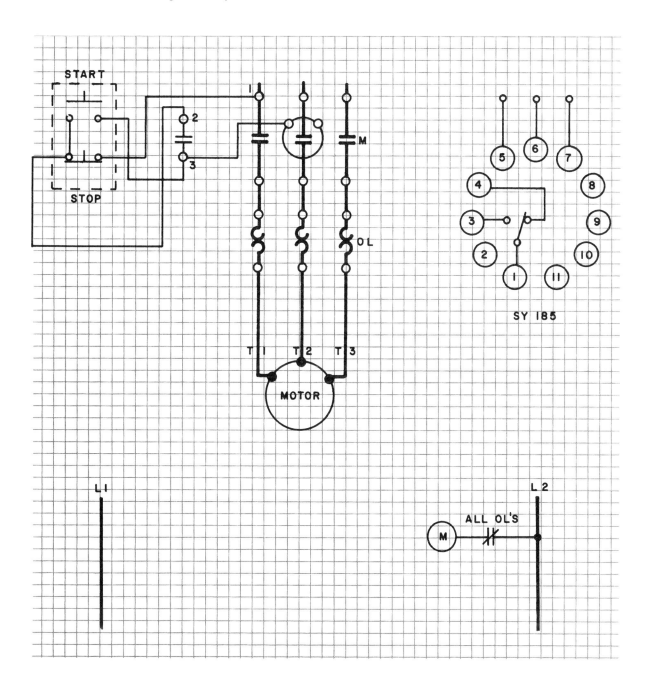

WORKSHEET 17-2

Score _____

Name _____ Class _____ Date _____

Referring to Data Sheet O, complete the wiring diagram and line diagram below, showing how this monitor (SM 170) can be used to de-energize the relay contacts if one or more phases is lost. This monitor will also de-energize the relay contacts if the phase sequence is not correct. In other words, L1 (called R) must be connected to pin 5; L2 (called S) must be connected to pin 6; and L3 (called T) must be connected to pin 7. Any other sequence will de-energize the relay.

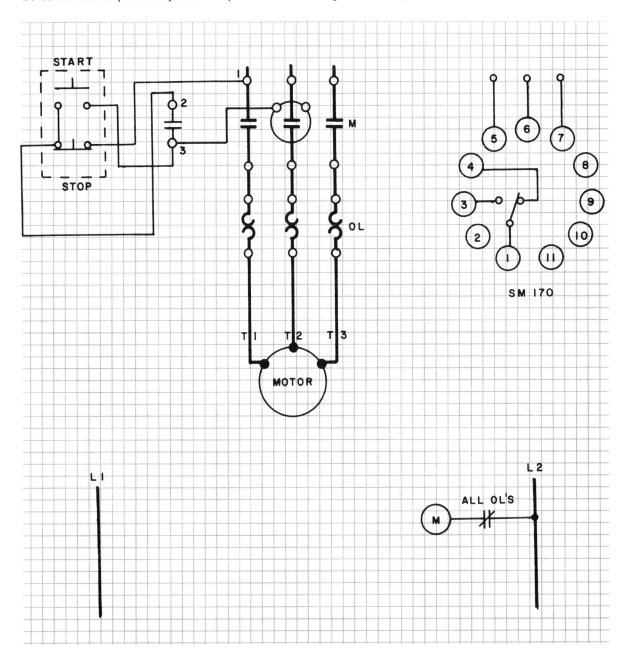

WORKSHEET 17-3

Score _____

Name _____ Class _____ Date _____

Referring to Data Sheet P, complete the wiring diagram and line diagram below, showing how this monitor (SM 115) can be used to protect against an overload in the control circuit or power circuit. This monitor can be used on single-phase or three-phase circuits. It uses a current transformer that detects the amount of current in any wire that passes through the current transformer. It can be used with or in place of the standard overload heater found on magnetic motor starters. The exact setting of the monitor, like the selection of overload heaters, must meet National Electrical Code requirements. Usually, the maximum set point of the monitor will be 1.15 or 1.25 times the FLC which is found on the nameplate of the motor. For exact requirements, see Article 430 of the National Electrical Code.

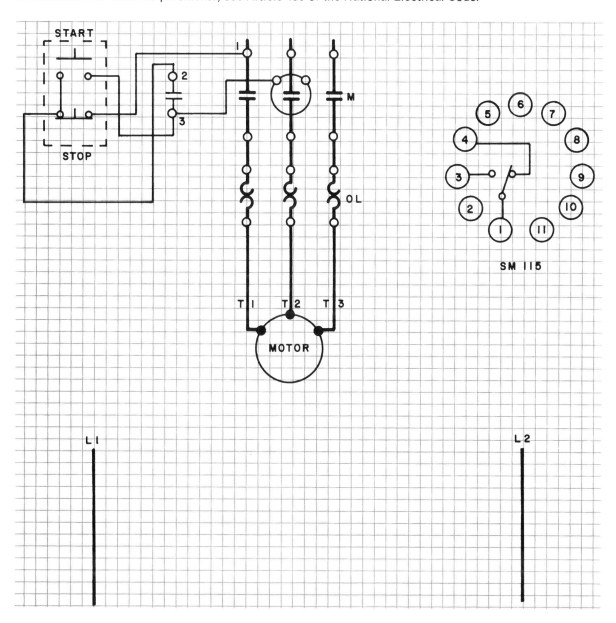

SM 115

WORKSHEET 17-4

Name _____ Class _____ Date _____

Because the SM 115 monitor on Worksheet 17-3 will react instantly to an overload in current, it will not allow the LRC (locked rotor current) required to start the motor. The relay will instantly detect the LRC and turn off the motor. Redraw the line diagram of Worksheet 17-3, adding a timer to allow the LRC to exist for five seconds before the SM 115 monitors the FLC (full load current).

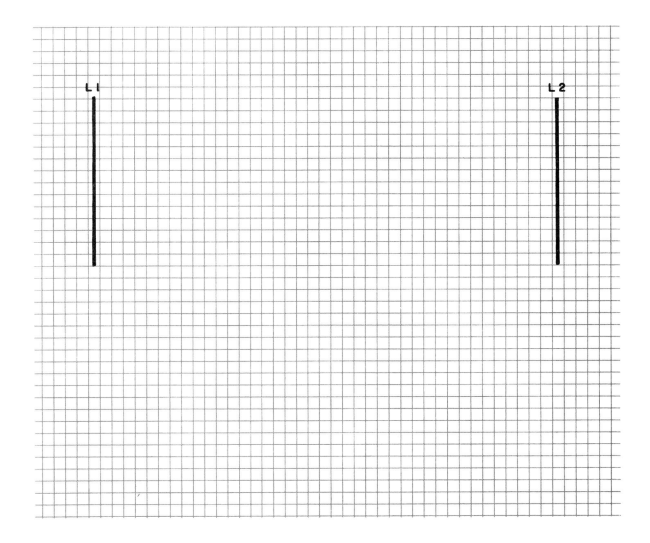

WORKSHEET 17-5

Name _____ Class _____ Date _____

Troubleshoot the circuit shown below according to the following information.

Problem: Motor does not turn on. Assume that the motor has been checked and is good. Both the float switch and overload contacts have been jumped with a fuse jump wire and found to be good.

Procedure: Circle with a red pencil the part of the circuit that probably contains the fault, both in the line diagram and wiring diagram.

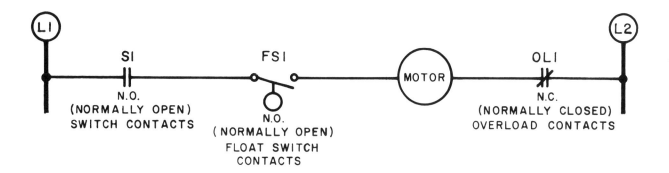

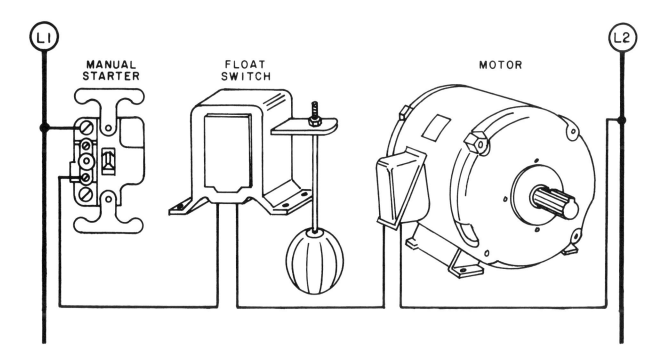

WORKSHEET 17-6

Name _____ Class _____ Date _____

Troubleshoot the circuit shown below according to the following information.

Problem: The solenoid energizes when the pushbutton or pressure switch is closed, but the pilot light does not come on when the foot switch and temperature switch close.

Procedure: (1) Illustrate where a fused jumper wire could be connected to eliminate trouble with the control switches. (2) Assume that the light still will not light when your jumper wire is in place. Circle with a red pencil the part of the circuit that probably contains the fault.

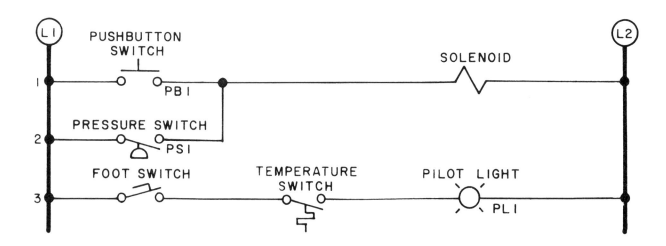

WORKSHEET 17-7

Name _____   Class _____   Date _____

Troubleshoot the circuit shown below according to the following information.

Problem: Although the empty boxes are stopping as they hit the limit switch and are staying in place for the given amount of time, they are not being filled.

Procedure: (1) Illustrate how a voltmeter could be connected into the circuit to test the solenoid valve. (2) Assume that the voltmeter indicates a proper voltage reading at the correct time. Circle with a red pencil the part of the circuit that probably contains the fault.

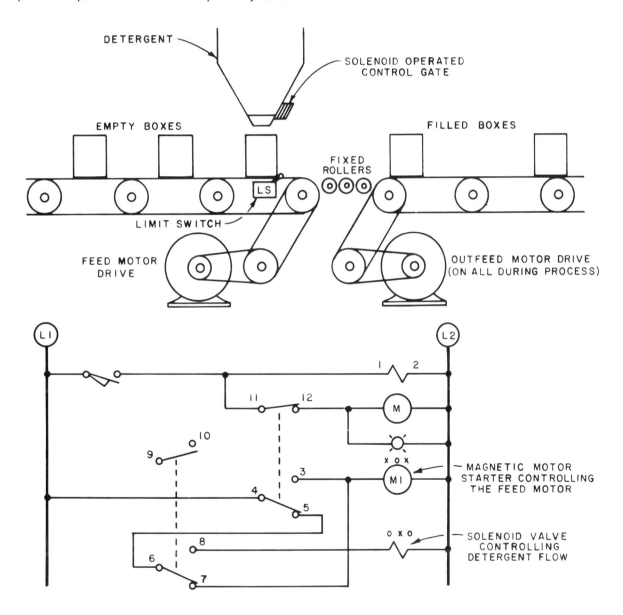

WORKSHEET 17-8

Score _____

Name _____ Class _____ Date _____

Troubleshoot the circuit shown below according to the following information.

Problem: Warning light PL2 is on. A check of starting coil M2 indicates that the starter is energized and the motor is running.

Procedure: Circle with red pencil that part of the circuit that probably contains the fault.

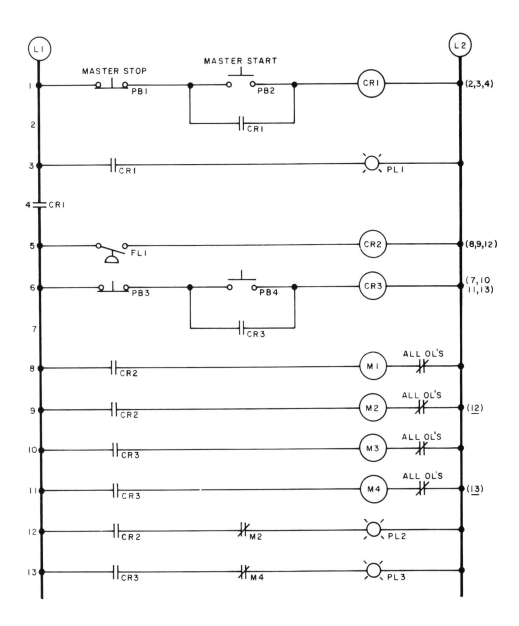

WORKSHEET 17-9

Name _____   Class _____   Date _____

Troubleshoot the circuit shown below according to the following information.

Problem: The motor is running hot and does not seem to have much power. A test with a voltmeter indicates that there is only about one-half the required voltage at terminals T1, T2, and T3 of the motor.

Procedure: (1) Illustrate how a fused jumper or jumpers could be connected to eliminate trouble with the control circuit. (2) Illustrate how a voltmeter could be connected to test the power circuit for the source of trouble.

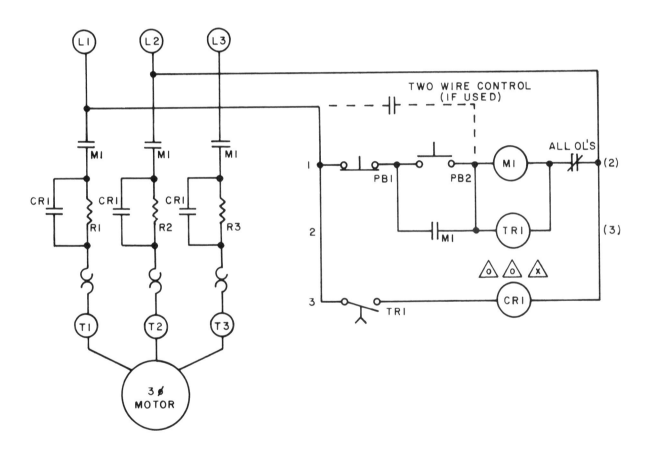

WORKSHEET 17-10

Score _____

Name _____ Class _____ Date _____

Troubleshoot the circuit shown below according to the following information

Problem: The motor is not braking to a stop. A test of the brake contactor in the control circuit indicates that the contactor is energizing for the correct time.

Procedure: (1) Illustrate where the AC voltmeter could be connected to test the brake contactor and transformer in the power circuit. (2) Illustrate where the DC voltmeter could be connected to test for a DC output.

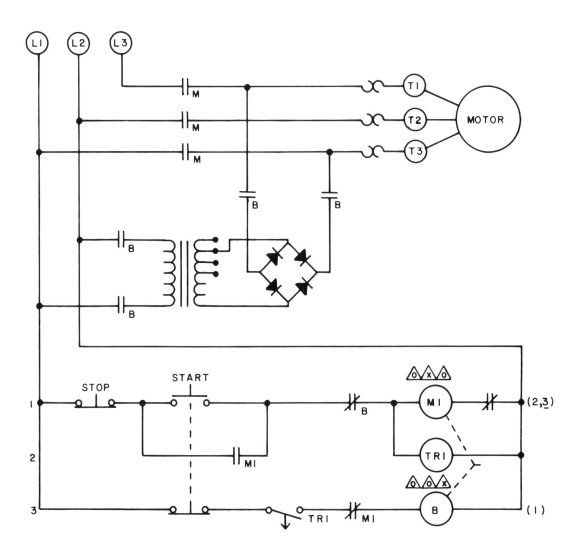

WORKSHEET 17-11

Name _____ Class _____ Date _____

Troubleshoot the circuit shown below according to the following information.

Problem: When the HIGH pushbutton is pressed, the motor starts in low, but after the set time the high motor starter will not come on, even though the low motor starter drops out.

Procedure: Circle with a red pencil the part of the circuit that probably contains the fault.

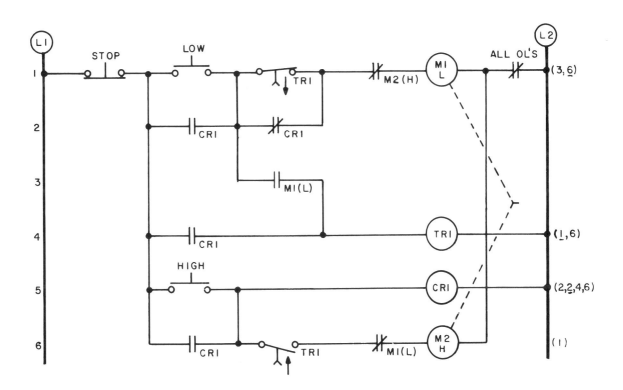

WORKSHEET 17-12

Score _____

Name _____ Class _____ Date _____

Troubleshoot the circuit shown below according to the following information

Problem: Magnetic starter coil M1 starts and remains engaged after the start pushbutton is pressed, regardless of the position of the selector switch.

Procedure: Circle with red pencil the part of the circuit that probably contains the fault.

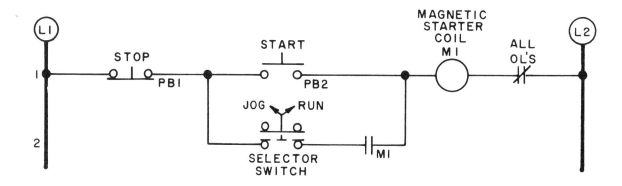

TECH-CHEK ✓ 17

Name _____

Class _____ Date _____

Score _____

**Preventive Maintenance and Troubleshooting
Techniques and Applications**

Select the best way to complete each statement and circle a, b, c, or d.

1. Phase unbalance in a three-phase power system usually occurs because
   a. three-phase loads are removed from the system.
   b. three-phase loads are added to the system.
   c. single-phase loads are added to the system.
   d. single-phase loads are removed from the system.

2. A kind of motor that will continue to run with a phase loss is the
   a. single-phase motor.
   b. three-phase motor.
   c. direct-current motor.
   d. capacitor-run motor.

3. On all equipment transporting people, the National Electrical Code requires protection against
   a. phase loss.
   b. phase reversal.
   c. phase unbalance.
   d. single-phasing.

4. Always remove the component to be tested or disconnect the line voltage from the circuit when making a
   a. resistance measurement.
   b. current measurement.
   c. voltage measurement.
   d. power measurement.

5. When using a fused jumper for troubleshooting, be sure *never* to connect the jumper between
   a. L1 and L2.
   b. L1 and the switch side of the load.
   c. L1 and the control switch.
   d. two control switches.

6. Each transformer is capable of delivering a limited current output at a given voltage. The power limit of a transformer is listed on the nameplate of the transformer under the
   a. voltage rating.
   b. current rating.
   c. secondary rating.
   d. KVA rating.

7. The major cause of most motor failures is
   a. capacitor failures.
   b. mechanical breakage.
   c. bearing failures.
   d. the deterioration of the insulation.

8. To indicate a good capacitor, the needle on an ohmmeter should
   a. swing to zero and not move.
   b. remain on infinity.
   c. swing to zero and slowly move one-fourth to one-third to one-half of the way across the scale to infinity.
   d. swing to infinity and slowly move one-half to one-third to one-fourth of the way across the scale to zero.

Continued

201

9. To indicate a good diode, an ohmmeter should show
   a. low resistance in one direction and high resistance in the other direction.
   b. low resistance in both directions.
   c. high resistance in both directions.
   d. middle-of-the-scale resistance in both directions.

10. On a ohmmeter, a zero reading in testing a fuse indicates that the fuse is
    a. good.
    b. bad.

11. In a DC motor, the series-field will usually have a reading
    a. more than the armature reading.
    b. less than the armature reading.

12. In a DC motor the shunt-field will usually have a reading
    a. more than the armature reading.
    b. less than the armature reading.

Fill in the blanks to complete each statement.

13. In using any instrument that has several scales, always start with the

    _____ scale available.

14. The voltage on incoming power should be within _____ percent of the

    voltage rating which appears on the nameplate of the loads that are connected to the lines.

15. When checking an SCR in the circuit with a clamp-on ammeter, you will probably find a reading

    of about _____ percent less than the actual current drawn in the circuit.

16. A _____ can be used to test an SCR that has been removed from

    the circuit.

17. To test the amount of work a motor is delivering, the reading from a _____

    can be compared to the given rating on the nameplate of the motor.

18-19. When testing for a Wye or Delta motor, remember that the Wye nine-lead motor will have

    (18)_____ separate circuits, and the Delta nine-lead motor will have

    (19)_____ separate circuits.

DATA SHEET A

Figure 1 shows a typical dual-element heating coil, like those found on most electric ranges. Figure 2 is a schematic diagram of the two heating elements. With these two elements and a 120/240-volt power supply, it is possible to have as many as ten different temperature ranges, depending on how you connect the elements and power supply together. Figures 3 and 4 show two possible switching arrangements. Connections to the power supply can be made in series, in parallel, or in series/parallel combinations.

Figure 1

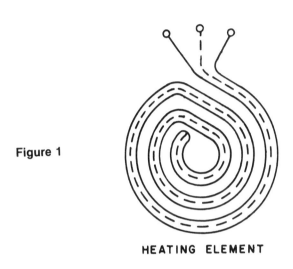

HEATING ELEMENT

Figure 2

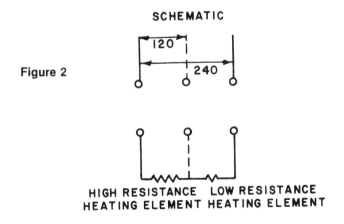

SCHEMATIC

HIGH RESISTANCE   LOW RESISTANCE
HEATING ELEMENT HEATING ELEMENT

Continued

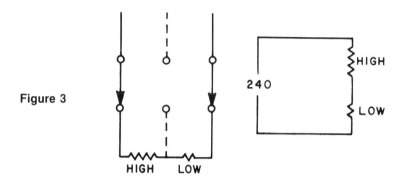

Figure 3

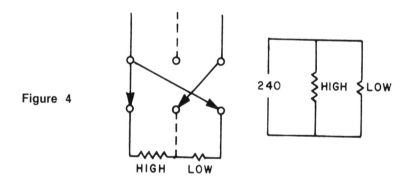

Figure 4

DATA SHEET B

MARKING WIRE NUMBERS ON LINE AND WIRING DIAGRAMS

A control circuit can consist of a few individual wires or of several thousand interconnected wires. Tracing one of these wires could be impossible without a system for keeping track of each wire in the circuit. The standard industrial system is to give each wire or *common group* of wires a number. A common group is defined as any set of wires that are connected directly without being broken by any device such as a pushbutton, contact, or starting coil.

As shown in Figure 1, numbers are assigned in numerical order (1, 2, 3, etc.), beginning from the upper lefthand corner and moving to the right, line by line. A new number is assigned whenever a wire (or common group of wires) is broken by an electrical device.

In Figure 1, the first wire is assigned number 1 and is called Wire Number 1 until it is broken by the stop pushbutton. After the stop pushbutton, Wire Number 2 begins and continues until it is broken by the start pushbutton and the M contact. Both the wire going to the start pushbutton and the wire going to the M contact are called Wire Number 2, because Wire Number 2 has not been broken yet by an electrical device. After the start pushbutton and after the M contact, both wires are called Wire Number 3 until broken by another electrical device — the M starting coil. After the starting coil, number 4 is assigned, and Wire Number 4 continues until it is broken by the overload contact. Then Wire Number 5 continues to the end of the circuit.

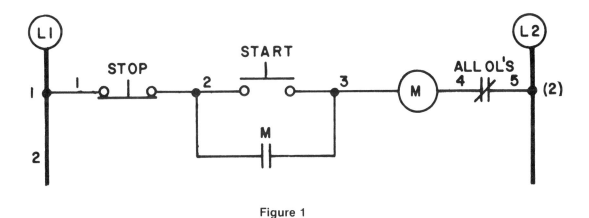

Figure 1

Figure 2 (on the other side of the page) further illustrates how wires in circuits are numbered. In this circuit seven numbers are assigned, some to individual wires and some to common groups of wires. A new number is assigned each time the interconnected wires are broken by an electrical device.

Continued

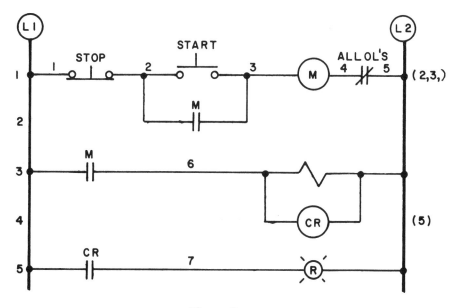

Figure 2

SIMPLIFYING THE WIRE NUMBERING SYSTEM

In Figures 1 and 2 each wire was assigned a number regardless of its purpose or location in the circuit. Although this practice is followed for some circuits, for most circuits the numbering system is simplified by not numbering wires that are usually prewired by the manufacturer prior to shipping. An example of prewiring is the wire connecting the motor starting coil to the overload contact in Figure 3. This wire need not be assigned a number.

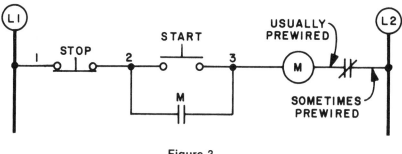

Figure 3

Sometimes the numbering system is further simplified by not numbering the wire directly connected to L2. This is because L2 to the overload contact is often prewired. The first wire from L1, however, is always numbered.

Continued

It is important to remember that the numbering system used for a given circuit applies to that circuit only. Numbering can differ from one circuit to another even if the circuits are electrically the same. An example of three circuits that are electrically the same but are numbered differently is shown in Figure 4.

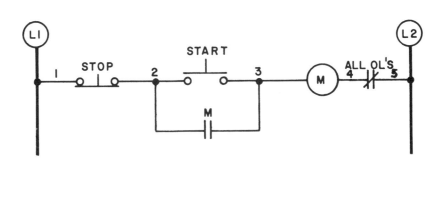

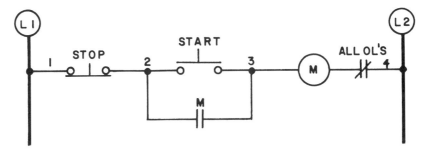

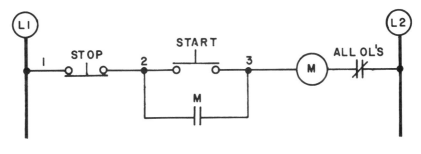

Figure 4

THE ADVANTAGES OF A WIRE NUMBERING SYSTEM

The use of a wire numbering system simplifies both the initial wiring and troubleshooting later. The need for such a system in wiring may not be apparent to you when you are working with a line diagram, but you will see its importance when you work with a wiring diagram in which wires enter and exit the conduit and are often grouped with many other wires.

Continued

Another advantage of a wire numbering system is that its use by manufacturers allows them to provide illustrations of connections for different wiring combinations. Figure 5 is a typical wiring diagram provided by a manufacturer of motor starters. When a motor starter is purchased with an enclosure, the manufacturer will include the diagram on the inside cover of the enclosure. All major manufacturers use the standard wire numbering system described in this Data Sheet with only slight variations.

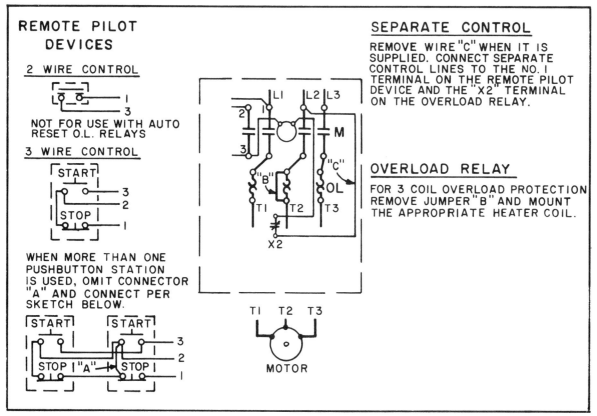

Figure 5

As illustrated in Figure 5, the manufacturer has prewired the motor coil to one side of the overload contact. The other side of the overload contact is connected directly to L2 and is marked X2. The note under the heading SEPARATE CONTROL explains why the contact is marked X2.

If the starting coil is the same voltage level as the supply lines (L1 and L2), the prewiring is left alone. However, if the starting coil is not the same voltage level as the supply lines, or if a separate control supply is used, wire C is removed from L2. An example of this would be if a 120-volt starting coil controlled a 240-volt motor. In that case L1 and L2 would have to be 240 volts in order to power the motor, and wire C would have to be removed and connected to a 120-volt supply. It is for this reason that X2 is marked on the overload, indicating a connection to a lower level supply voltage.

Notice in Figure 5 that the manufacturer has marked wire points 1, 2, and 3 on the diagram. These numbers are standard and are marked directly on all major starters. With each of these points marked in this manner, it is very easy to wire this starter for two-wire, three-wire, or multiple-control stations. If, for example, a two-wire control device such as a pressure float or temperature switch were used to control the motor, the control device would be connected to points 1 and 3. If a three-wire control device, such as two pushbuttons, were used to control the motor, it would be connected as shown in Figure 5.

Unfortunately, as circuits become more complicated, it is impossible for the manufacturer to apply a standard numbering system to them. Instead, the electrician needs to number the wires in each individual circuit at the time of wiring. An exception are those circuits that are very frequently used, such as those for forwarding and reversing a motor with a standard STOP-START station, in which case the manufacturer can provide the numbering.

USING INFORMATION FROM LINE DIAGRAMS TO CONNECT WIRING DIAGRAMS

The basic language of controls is the line diagram. Its function is to illustrate quickly and concisely how the control circuit is to perform electrically. With the use of the line diagram the wiring diagram can be completed. Unlike the line diagram, the wiring diagram shows as closely as possible the actual connections and placement of all components in the circuit.

Figure 6 is the line diagram of a standard START/STOP pushbutton station with MEMORY. Figure 7 is the wiring diagram for the same station. Note that the wiring diagram shows all connections of all components, even those that are present but not used in this particular circuit — for example, the NC start contacts, NO stop contact, and the NC auxiliary contacts on the starter.

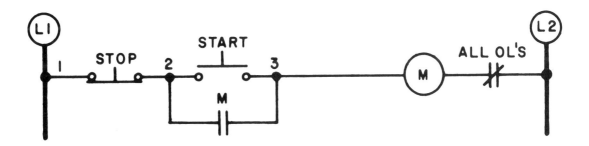

Figure 6

Continued

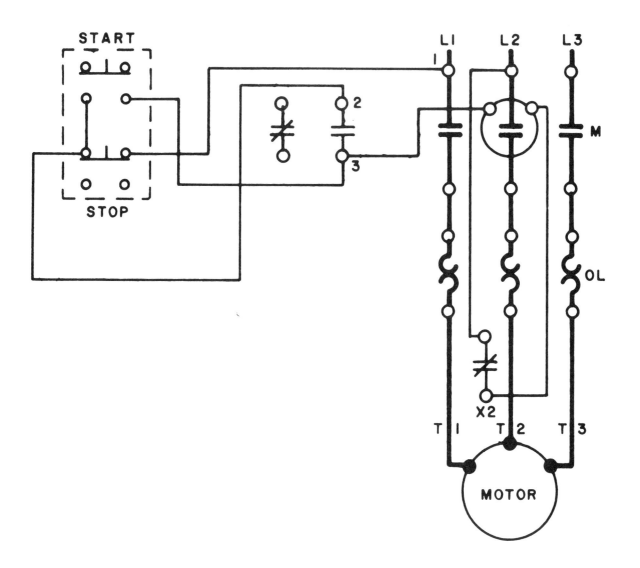

Figure 7

DATA SHEET C

Figure 1 is a pictorial view and Figure 2 is a line diagram of a system for controlling the back and forth motion of a fluid power cylinder. Also illustrated are the conduit connections for each piece of electrical equipment.

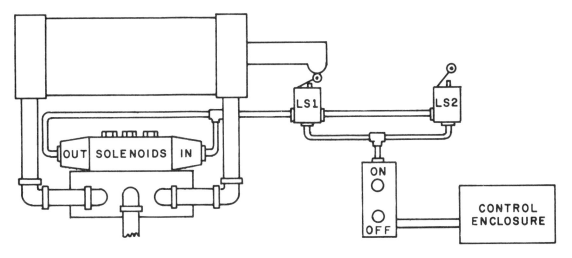

Figure 1

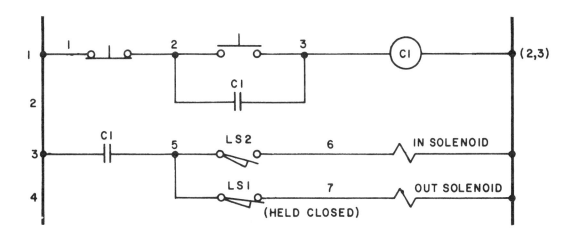

Figure 2

Continued

DATA SHEET D

USING A CODING TABLE

A coding system can be added to the loads in timing circuits to determine the condition of the load for each sequence of the circuit operation. For example, a code of reset, timing, and timed out can be added to a circuit that uses an ON-delay timer. This coding system is useful in understanding the control circuit and can also be used to design circuits that use timers.

If we take the typical ON-delay timer used in a circuit that includes the standard sequence of reset, timing, and timed out, we can determine that only eight possibilities exist for any output load. These eight possibilities are listed in Figure 1.

	Reset	Timing	Timed Out
1	O	O	O
2	O	O	X
3	O	X	O
4	O	X	X
5	X	O	O
6	X	O	X
7	X	X	O
8	X	X	X

O = load de-energized
X = load energized

Figure 1

The table in Figure 1 can be simplified by eliminating the sequences of OOO and XXX, since they do not represent a useful control function. In the case of OOO, the load would be de-energized all the time, and in the case of XXX the load would be energized all the time. The other six possibilities provide control sequences that are useful and can be accomplished through the contacts normally provided on a timer.

COUNTING A CODING TABLE IN CONTACT ARRANGEMENTS

Timers are usually provided with NO and NC time-delay and instantaneous contacts. By using these timer contacts in various configurations, we can associate certain contact arrangements directly with the remaining six sequence codes in Figure 1.

For example, if a NO time-delay contact were connected in series with a load, as illustrated in Figure 2, the code of OOX would result.

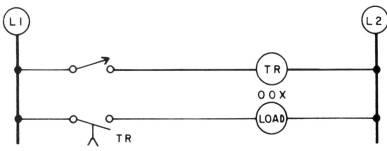

Figure 2

Continued

212

The implication of this statement is that any NO time-delay contact in series with a load will provide the necessary sequence of OOX. Thus the electrician can develop a simple table, as shown in Figure 3, with all six possible combinations, that would immediately tell him which contacts to use and how to connect them for any given sequence.

NO TIME DELAY	NO INSTANTANEOUS
NC TIME DELAY	NC INSTANTANEOUS

SEQUENCE CODE	CIRCUIT CONNECTION
O O X	
O X O	
O X X	
X O O	
X X O	
X O X	

Figure 3

DATA SHEET E

Once you understand how the simple table in Figure 3 in Data Sheet D was developed, you are ready to begin designing basic timing circuits, using the information in that table.

The first step in designing a circuit is to list each load according to its requirements in the circuit. The second step is to list the control devices (such as a pushbutton) that will be needed. For example, assume that you are asked to develop a circuit to control the length of time a cake is to be baked in an oven and to indicate by a bell when it is to be removed from the oven. First, you can see that there are two loads to be controlled: the oven and the bell. Since the bell will signal the end of timing, the code for the bell will be OOX. Since the oven will bake for a predetermined length of time, the code for the oven will be OXO. From this information, and using the table developed in Data Sheet D (which lists each code combination) you can design the first part of the circuit, as shown in Figure 1.

STEP I

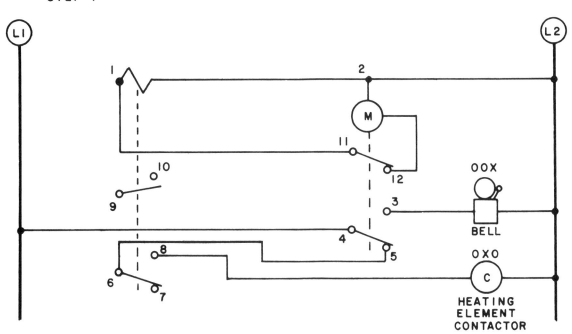

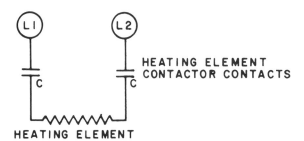

HEATING ELEMENT

Figure 1

Continued

The next step in completing the circuit is to add the control devices necessary to control the timer and loads. Two controls are required: one to signal the start and end of the baking process and another to control the oven temperature. A simple on-off switch can be used to signal the start and end of the process, and a temperature switch can control oven temperature. Adding these two controls to the timing circuit makes the circuit complete, as illustrated in Figure 2.

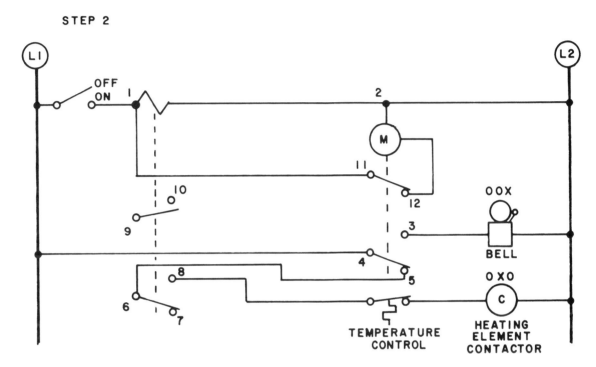

Figure 2

SV 115/215
Adjustable sensitivity

DATA SHEET F

SV 115/215

Courtesy of Electromatic Controls Corp.
2495 Pembroke Ave.
Hoffman Estates, Illinois

❄ **Level control for conductive liquids.**
❄ **MAX.-MIN. control of DISCHARGING.**
❄ **Adjustable sensitivity.**
❄ **10 A SPDT or 5 A DPDT output relay.**
❄ **LED-indication for relay on.**
❄ **AC- or DC supply voltage.**

SPECIFICATIONS

Sensitivity
Knob-adjustable sensitivity
with relative scale.
ON from 3.5 KΩ to 25 KΩ.
OFF from 8 KΩ to 45 KΩ.

Sensor voltage
Max. 24 VAC.

Sensor current
Max. 2.5 mA.

**Connection cable
between sensor and
amplifier**
2- or 3-core plastic cable,
normally unscreened.
Cable length:
Max. 100 metres.

The resistance between the
cores and ground must be at
least 220 KΩ.
In certain cases it is recommended to use screened cable
between sensor and amplifier,
e. g. where the cable is placed
parallel to load cables.
The screen is connected to
pin 7.

Accessories
Bases.
Hold down spring.
Mounting rack.
Base cover.
Front mounting bezel.
Sensors type VH - VN -
VT and VS.
Nut type VM.1,5.

WIRING DIAGRAMS

Example 1

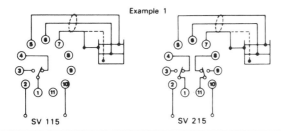

SV 115   SV 215

Example 2

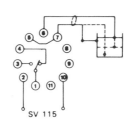   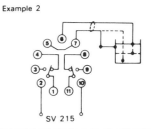

SV 115   SV 215

MODE OF OPERATION

Max. and/or min. control for DISCHAR-
GING of conductive liquids.
Relay for control of CHARGING, see
SV 125/225 page 9.

Example 1
The diagram shows the level control
connected as max. and min. control, i.e.
registration of 2 levels.
The relay operates when the liquid reaches the max. electrode (pin 5), provided
that the min. electrode (pin 6) is in contact with the liquid.

The relay releases when the min. electrode is no longer in contact with the liquid. Pin 7 has to be connected to the
container. If the container consists of a
non-conductive material, an additional
electrode has to be used which is connected to pin 7.
In the diagram this electrode is shown
by the dotted line.

Example 2
The diagram shows the level control
connected as max. or min. control, i.e.
registration of 1 level.
The relay operates when the electrode
(pin 6) is in contact with the liquid.
Again an additional electrode has to be
used, if the container consists of a non-
conductive material.
(To be connected to pin 7).

OPERATION DIAGRAM Ex. 1.

| Supply voltage |
| Max. electrode (pin 5) in liquid |
| Min. electrode (pin 6) in liquid |
| Relay on |

OPERATION DIAGRAM Ex. 2.

| Supply voltage |
| Min. electrode (pin 6) in liquid |
| Relay on |

DATA SHEET G

Electricians must be able to wire a variety of AC/DC motors to run in forward and reverse. Wiring these motors can be confusing when the motor diagram is new or unfamiliar to you. To add to this confusion, almost every motor manufacturer provides only one basic diagram. Modifications such as reversing are listed as printed information below the wiring diagram. Thus it becomes the electrician's job to convert the wiring diagram and written instructions into a circuit that will properly reverse the motor.

In this Data Sheet a basic wiring procedure is explained step by step. Once you understand this basic procedure, you can modify it to develop the proper circuitry to wire any AC or DC motor to run in forward and reverse at any voltage. As you follow this basic procedure, be sure you understand each example given before you move on to the next. Although this procedure will seem long the first time through, with practice you will be able to simplify it. After applying it to several situations, you will find that you can develop a wiring diagram for any motor in just a few minutes.

BASIC WIRING RULES

Before you begin to learn this basic wiring procedure, you should first be familiar with the following basic rules.

1. *Every "hot" lead must be switched.* The "hot" lead (also called the ungrounded conductor) must be switched so that power will never be applied to the motor in the OFF position. Figure 1 shows each of the basic motor types with the "hot" lead switched.

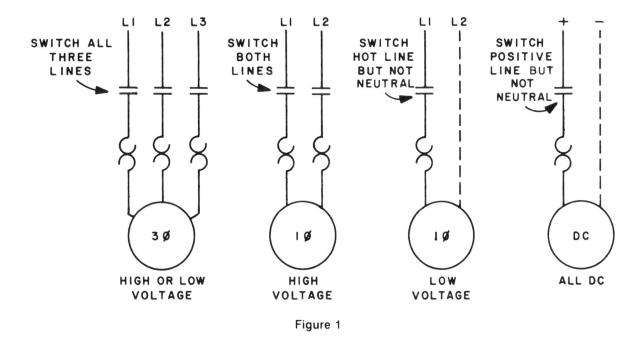

Figure 1

2. *A motor must be connected for only one voltage.* Motors are often designed with multiple windings for different voltages. Be sure to wire the motor for the proper voltage in each case.

Continued

217

3. *A motor must be connected so that it cannot be electrically told to run simultaneously in forward and reverse directions.* Obviously, your wiring diagram must provide for the motor to run in only one direction at a time. Otherwise, the motor or the power source could be severely damaged.

4. *A motor must be connected so that it cannot be electrically told to run simultaneously at different speeds.* For a multi-speed motor your wiring diagram must provide for the motor to run at only one speed at a time. Otherwise, the motor or the power source could be severely damaged.

STEP-BY-STEP BASIC WIRING PROCEDURE

The following seven basic steps are necessary for wiring any AC or DC motor to run in forward and reverse.

1. *Develop a motor objective.* You must first determine exactly what your wiring diagram is to accomplish. As a result, you will select only one of the two voltage connections illustrated on the motor. For our first example, (Example 1), assume that you want to connect a motor so that it will run in forward or reverse by means of a magnetic controller.

2. *Obtain information from the motor nameplate.* The motor nameplate provides information for two voltage supplies. Figure 2 shows the nameplate information on a typical three-phase motor which can be wired for 480 volts or 220 volts. For Example 1, assume that the supply voltage is 220. Consequently, you will use the information provided for low voltage, as shown in Figure 3.

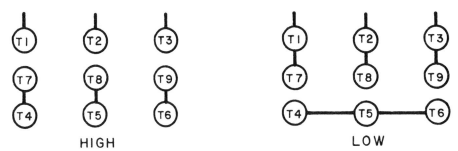

HIGH   LOW

Figure 2

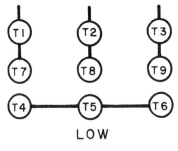

LOW

Figure 3

Continued

3. *Make a written diagram of your objective, showing exactly how each wire must be connected to accomplish your objective.* Simply put the wiring connections into words, as shown in Figure 4. In this example the manufacturer's wiring diagram has been followed, with L1 and L3 interchanged as specified by NEMA standards.

Forward	Reverse
L1 to T1 & T7	L1 to T3 & T9
L2 to T2 & T8	L2 to T2 & T8
L3 to T3 & T9	L3 to T1 & T7
T4 to T5 to T6	T4 to T5 to T6

Figure 4

4. *Remove common connections that are not power lines.* To simplify your written diagram (or objective), you can remove any connection that is the same in both forward and reverse, provided it is not power line. Figure 5 shows that connections T4, T5, and T6 in our example can be removed, because they are the same in forward and reverse. Connections L2, T2, and T8 cannot be removed, because L2 is a power line and must be switched. With T4, T5, and T6 removed, the objective will look like Figure 6.

T4 to T5 to T6 (can be removed)

L2 to T2 & T8 (must not be removed)

Figure 5

Forward	Reverse
L1 to T1 & T7	L1 to T3 & T9
L2 to T2 & T8	L2 to T2 & T8
L3 to T3 & T9	L3 to T1 & T7

Figure 6

5. *Use one name for remaining common connections.* To further simplify your written objective, you can use one name for any combination of wires that appears in the same combination on both sides of the objective (that is, for both forward and reverse). Figure 7 shows which combinations in our example can be given one name. Figure 8 shows the simplified objective that results.

T1 & T7 = T1
T2 & T8 = T2
T3 & T9 = T3

Figure 7

Forward	Reverse
L1 to T1	L1 to T3
L2 to T2	L2 to T2
L3 to T3	L3 to T1

Figure 8

Continued

6. *Convert the written diagram of your objective into a diagram showing the placement of electrical contacts.* Once you have a simplified written diagram of your objective, the wiring diagram is very easy to determine. Every place the word "to" appears in your written objective, a set of electrical contacts will be needed. Figure 9 shows how our objective for Example 1 looks with the symbol for electrical contacts replacing the word "to".

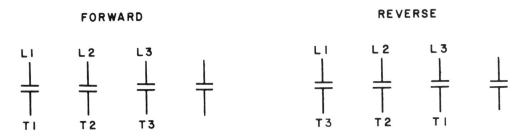

Figure 9

7. *Draw the wiring diagram.* To draw your wiring diagram, you need only to show connections between any two motor or power lines that are the same. In our example, in Figure 10, L1 on the forward side is connected to L1 on the reverse side. Likewise, T1 on the forward side is connected to T1 on the reverse side. When all the lines are drawn, your wiring diagram is complete.

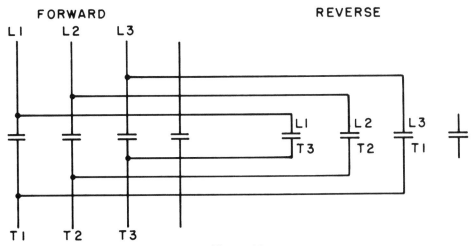

Figure 10

FURTHER EXAMPLES USING STEP-BY-STEP PROCEDURE

For Example 2 we will use the nameplate information for a typical single-phase motor that can be wired for 115 volts or 230 volts, shown in Figure 11.

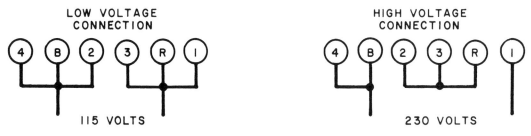

Figure 11

Continued

Again, let us follow the steps in the basic wiring procedure.

1. *Develop a motor objective.* In Example 2 we want to connect a motor to run in forward or reverse by means of a manual controller.

2. *Obtain information from the motor nameplate.* Assume a supply voltage of 115 for Example 2, and obtain the appropriate information from Figure 11.

3. *Make a written diagram of your objective.* The written objective for Example is shown in Figure 12. Please note that it has not yet been simplified. It is important that you do not combine steps or simplify your objective too early. This is the cause of most mistakes. Remember that power lines must be switched; therefore, each must be listed separately. Note that the two wires to be interchanged in this example are black and red (B and R). Consequently, they have to be listed separately and not connected to any other wire.

Forward	Reverse
B to 4 & 2	R to 4 & 2
R to 3 & 1	B to 3 & 1
L1 to 4 & 2	L1 to 4 & 2
L2 to 3 & 1	L2 to 3 & 1

Figure 12

4. *Remove common connections that are not power lines.* Figure 13 shows how the objective will look after common connections that are not power lines are removed.

Forward	Reverse
B to 4 & 2	R to 4 & 2
R to 3 & 1	B to 3 & 1
L1 to 4 & 2	L1 to 4 & 2

Figure 13

5. *Use one name for remaining common connections.* Figure 14 shows how the objective will look after it is further simplified by following this step.

Forward	Reverse
B to 4	R to 4
R to 3	B to 3
L1 to 4	L1 to 4

Figure 14

Continued

6. *Convert the written objective to a diagram showing electrical contacts.* Figure 15 shows the result of following this step.

Figure 15

7. *Draw the wiring diagram.* Figure 16 shows the final wiring diagram for Example 2.

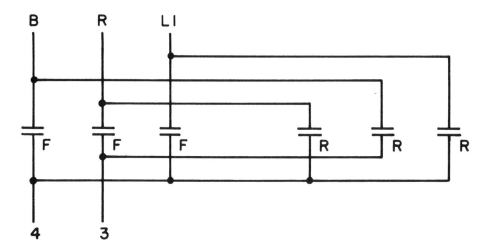

Figure 16

For Example 3 we will follow the step-by-step procedure for both voltage ratings listed for a single-phase, dual-voltage capacitor-start motor, as shown in Figure 17.

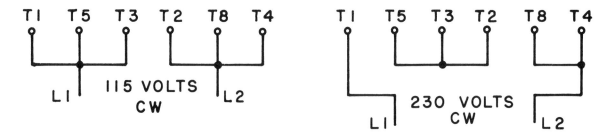

TO REVERSE, INTERCHANGE T5 & T8

Figure 17

Continued

222

1. *Develop a motor objective.* As for Example 2, our objective for Example 3 is to connect a motor to run in forward or reverse by means of a manual controller.

2. *Obtain information from the motor nameplate.* For Example 3, we will make two wiring diagrams — one for a supply voltage of 115 and one for a supply voltage of 230. For each, obtain the appropriate information from Figure 17.

3. *Make a written diagram of your objective.* The written objective for the 115-volt connection in Example 3 is shown in Figure 18. Note that T5 and T8 are listed separately because they are to be switched. Also, L1 and L2 are listed separately because they are power lines. Note, too, that no attempt has been made yet to simplify this objective.

Forward	Reverse
T5 to T1 & T3	T5 to T2 & T4
T8 to T2 & T4	T8 to T1 & T3
L1 to T1 & T3	L1 to T1 & T3
L2 to T2 to T4	L2 to T2 to T4

Figure 18

The objective for the 230-volt connection in Example 3 is shown in Figure 19.

Forward	Reverse
T5 to T2 & T3	T5 to T4
T8 to T4	T8 to T2 & T3
L1 to T1	L1 to T1
L2 to T4	L2 to T4

Figure 19

4 and 5. *Remove common connections that are not power lines, and use one name for remaining common connections.* Figure 20 shows the simplified objective for the 115-volt connection. T1 and T3 are to be wired together and called T1. T2 and T4 are to be wired together and called T2. Since L2 is a neutral wire on 115-volt circuits, it does not have to be switched (see Figure 1). L2 is to be connected to T2 and T4 and called T2.

Forward	Reverse
T5 to T1	T5 to T2
T8 to T2	T8 to T1
L1 to T1	L1 to T1

Figure 20

Continued

Figure 21 shows the simplified objective for the 230-volt connection. T2 and T3 are to be wired together and called T2. Neither L1 nor L2 can be directly connected, since both are "hot" wires on 230-volt circuits.

Forward	Reverse
T5 to T2	T5 to T4
T8 to T4	T8 to T2
L1 to T1	L1 to T1
L2 to T4	L2 to T4

Figure 21

6. *Convert the written objective to a diagram showing electrical contacts.* Figure 22 shows the result of following this step for the 115-volt connection, and Figure 23 shows the result for the 230-volt connection.

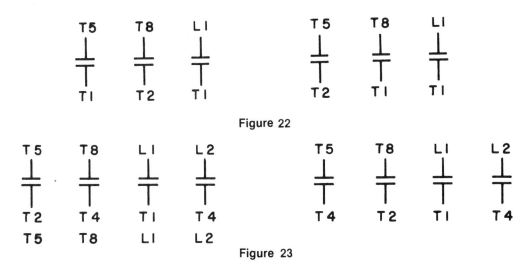

Figure 22

Figure 23

7. *Draw the wiring diagram.* Figure 24 shows the final wiring diagram for the 115-volt connection, and Figure 25 shows the final wiring diagram for the 230-volt connection. Figures 26 and 27 show the internal connections that change within the motor as a result of both types of wiring.

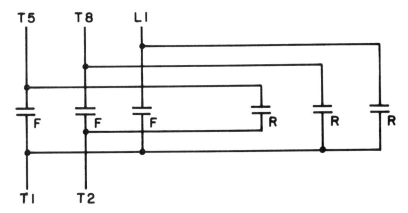

Figure 24

Continued

224

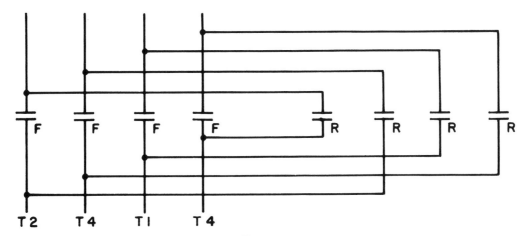

T 2   T 4   T 1   T 4

Figure 25

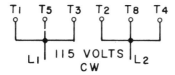

FOR CLOCKWISE ROTATION

T1   T5   T3   T2   T8   T4

L1   115 VOLTS   L2
CW

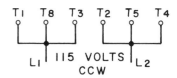

FOR COUNTERCLOCKWISE ROTATION

T1   T8   T3   T2   T5   T4

L1   115 VOLTS   L2
CCW

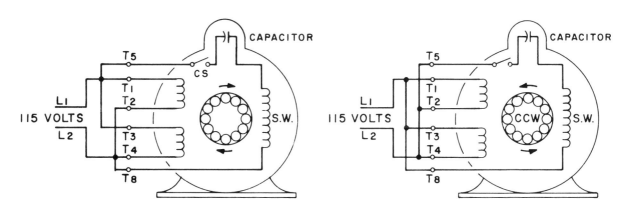

Figure 26

Continued

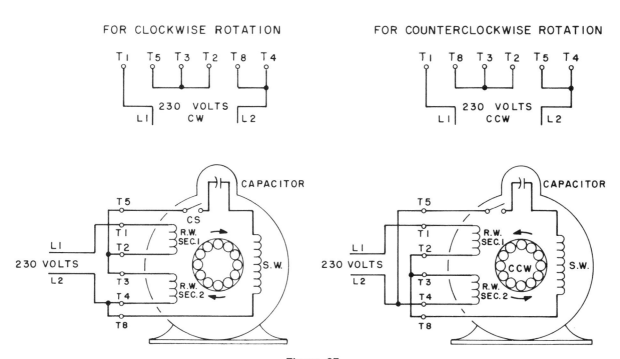

FOR CLOCKWISE ROTATION

T1 T5 T3 T2 T8 T4

230 VOLTS
L1   CW   L2

FOR COUNTERCLOCKWISE ROTATION

T1 T8 T3 T2 T5 T4

230 VOLTS
L1   CCW   L2

CAPACITOR

T5
CS
T1
R.W.
T2
SEC.1
L1
230 VOLTS
L2
S.W.
T3
R.W.
T4
SEC.2
T8

CAPACITOR

T5
T1
R.W.
T2
SEC.1
L1
230 VOLTS
L2
CCW
S.W.
T3
R.W.
T4
SEC.2
T8

Figure 27

DATA SHEET H

Figure 1 shows a small busway system composed of elbows, tees, crosses, feeder ducts, and plug-in ducts. Figure 2 (on the other side of the page) shows the sketch used to help make up a bill of materials for the job.

BILL OF MATERIALS

133 feet of feeder duct
47 feet of plug-in duct
1 tee
4 elbows
1 cross

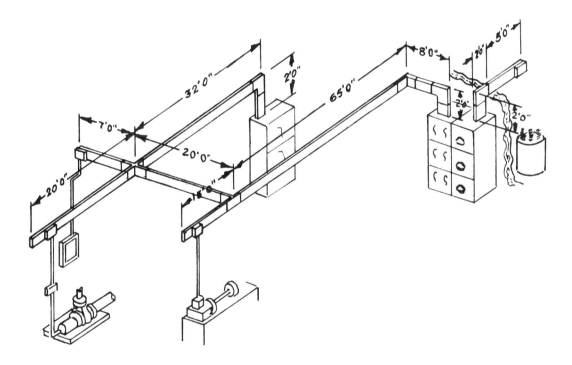

Continued

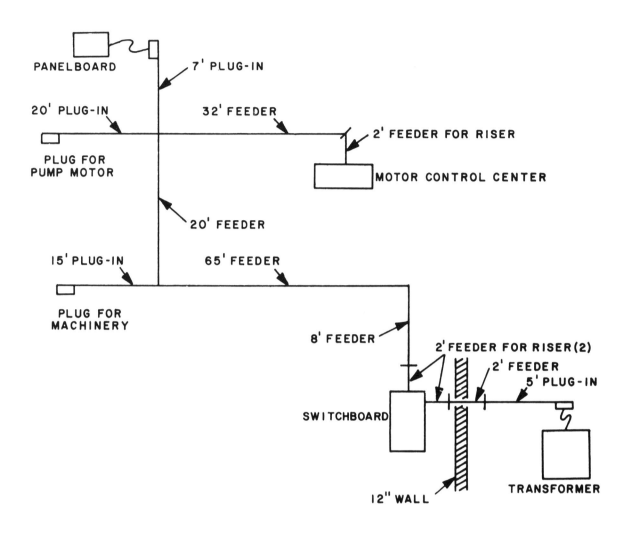

DATA SHEET I

SE 110

✳ **Relay for photosensors with <u>modulated</u>, infrared light.**
✳ **Built-in power supply for transmitter/-receiver.**
✳ **For separate transmitters and receivers with max. ranges: 1 to 100 metres.**
✳ **For combined transmitters and receivers with max. ranges: 1 to 10 metres.**
✳ **Transmitter- and receiver connections are short-circuit safe.**
✳ **10 A SPDT output relay.**
✳ **LED-indication for relay on.**
✳ **AC or DC supply voltage.**

SE 110
No adjustment

Courtesy of Electromatic Controls Corp.
2495 Pembroke Ave.
Hoffman Estates, Illinois

SPECIFICATIONS

Common technical data and ordering key
Pages 10 -12.

Frequency
Max. 10 pulses/s.

Duration of light/darkness
Both: Min. 50 ms.

Connections for transmitters
Voltage/current:
3,5 VDC - 100 mA.

Idle voltage:
5 VDC.

Short-circuit current:
500 mA.

Connection:
Pins 6 and 7.
Pin 7 positive.
Short-circuit safe.

Connections for receivers
Voltage: 12 VDC.

Current: Lit: 15 mA
Dark: 1 to 4 mA.

Idle voltage:
12 VDC.

Short-circuit current:
75 mA.

Connection:
Pins 5 and 6.
Pin 5 positive.
Short-circuit safe.

Accessories
Bases.
Hold down spring.
Mounting rack.
Base covers.
Front mounting bezel.

Infrared transmitters.
Infrared receivers.
Combined infrared transmitters and receivers.
Reflectors.
Separate power supply for special applications, type SE 010.
See catalogue on accessories.

WIRING DIAGRAMS

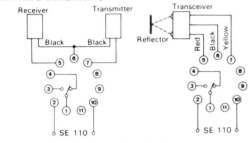
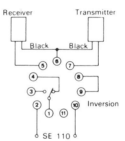

Example 1

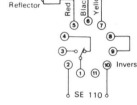

Example 2

MODE OF OPERATION

The relay is used in conjunction with separate, infrared transmitters and receivers and retro-reflective heads.
The photosensors work with infrared, modulated light and because of the modulation they are insensitive to environmental light.
The transmitter is a Ga-As diode and the receiver is a phototransistor.
When sensing by reflection either a reflector type ER or other materials can be used, such as plastics, textiles, metal, wood, paper, glass, etc.

Example 1: The relay releases, when the light beam is interrupted. The relay also releases in case one or more of the cables between the photosensors and the SE 110 are disconnected and in case of power failure.

Example 2: The relay operates, when the light beam is interrupted. The relay also operates in case one or more of the cables between the photosensors and the SE 110 are disconnected.

OPERATION DIAGRAM

Supply voltage			
Light beam interrupted			
Ex. 1: Relay on			
Ex. 2: Relay on			

DATA SHEET J

SD 110/210

* Relay for inductive and capacitive sensors without amplifier (NAMUR).
* Voltage- and current limitation in sensor circuit (8 VDC, 8 mA).
* Relay locks in »OFF« position by cable failures.
* 10 A SPDT or 5 A DPDT output relay.
* LED-indication for relay on.
* AC or DC supply voltage.

SD 110/210
No adjustment

Courtesy of Electromatic Controls Corp.
2495 Pembroke Ave.
Hoffman Estates, Illinois

SPECIFICATIONS

Commom technical data and ordering key
Pages 10–12.

Sensor voltage
Pins 5–6 or 6–7:
8 VDC.
Pin 6 positive.

Sensor current
Activated: < 1 mA
Not activated: > 3 mA

Short-circuit current
Max. 8 mA.

Connection cable
Unshielded PVC core.
Can be extended if required, maximum resistance:
100 Ω.

Sensing range
0.5–40 mm depending on the sensor.
See sensor specifications.

Sensing speed
Max. 10 operations/s.

Pulse time
Min. 20 ms.

Subject of detection
Solid, fluid, or granulated substances.

Accessories
Bases.
Hold down spring.
Mounting rack.
Base covers.
Front mounting bezel.

Inductive sensors, type DU, DJ and DO. Capacative sensors, type DR.
See catalogue on accessories.

WIRING DIAGRAM

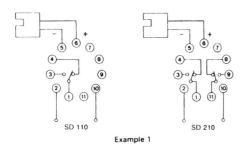

SD 110   SD 210

Example 1

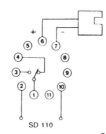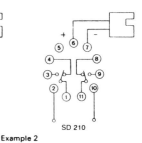

SD 110   SD 210

Example 2

MODE OF OPERATION

Example 1
The relay operates by activation of the sensor.
It releases automatically in case of cable failure.

Example 2
The relay releases by activation of the sensor.
It releases automatically in case of cable failure.

OPERATION DIAGRAM

Supply voltage

Sensor activated

Cable failure

Ex. 1: Relay on

Ex. 2: Relay on

DATA SHEET K

SD 170/270

SD 170/270
No adjustment

* **Bistable relay for 2 inductive or capacitive sensors without amplifier (NAMUR).**
* **Voltage- and current limitation in sensor circuits (8 VDC, 8 mA).**
* **10 A SPDT or 5 A DPDT output relay.**
* **LED-indication for relay on.**
* **AC or DC supply voltage.**

Courtesy of Electromatic Controls Corp.
2495 Pembroke Ave.
Hoffman Estates, Illinois

SPECIFICATIONS

Commom technical data and ordering key
Pages 10–12.

Sensor voltage
Pins 5–6 or 6–7:
8 VDC.
Pin 6 positive.

Sensor current
Activated: < 1 mA
Not activated: > 3 mA

Short-circuit current
Max. 8 mA.

Connection cable
Unshielded PVC core.
Can be extended if required,
maximum resistance:
100 Ω.

Sensing range
0.5–40 mm depending on the
sensor.
See sensor specifications.

Sensing speed
Max. 10 operations/s.

Pulse time
Min. 20 ms.

Subject of detection
Solid, fluid, or granulated
substances.

Accessories
Bases.
Hold down spring.
Mounting rack.
Base covers.
Front mounting bezel.

Inductive sensors, type DU,
DJ and DO.
Capacative sensors, type DR.
See catalogue on accessories.

WIRING DIAGRAM

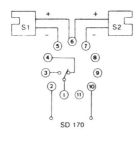

SD 170

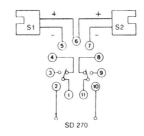

SD 270

MODE OF OPERATION

The SD 170/270, being bistable relays, are used with 2 proximity sensors in the following way:
The relay operates when one of the sensors (S 1) is activated momentarily and then remains operated.

When the other sensor (S 2) is activated momentarily, or when the supply voltage is interrupted, the relay releases.

If both sensors are activated at the same time, the relay releases, or shall not operate respectively.
Sensor S 2 has priority.

OPERATION DIAGRAM

Supply voltage	
Sensor S 1 activated	
Sensor S 2 activated	
Relay on	

DATA SHEET L

PROGRAMMING IN GENERAL

The programming of most programmable controllers use the language of line diagrams to write and enter the logic of the control circuit into the controller. This means that the sequence of commands should be as close to the normal line diagram as possible. The drawing below illustrates a typical line diagram and the required program needed to enter this logic into the programmable controller.

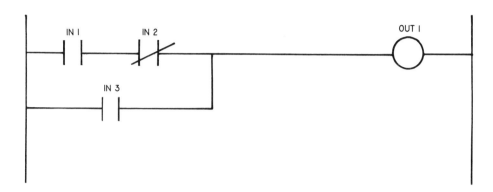

Memory location	INSTRUCTION		OPERAND		Title: date:
0	LD		IN	1	
1	AND	NOT	IN	2	
2	OR		IN	3	
3	=		OUT	1	
4	END				
5					
6					
7					
8					
9					

In the above circuit, IN 1, 2 or 3 can be any input such as a pushbutton, limit switch or photoelectric contact. Output 1 can be any output such as a light, motor starter or solenoid. When using programmable controller format, the following symbols are used.

—| |—

DESCRIBES A MAKE FUNCTION

—|/|—

DESCRIBES A BREAK FUNCTION

—◯—

DESCRIBES AN OUTPUT FUNCTION

When you build up a standard control circuit, the connections in series and in parallel are made so that you connect the individual components with wire. This is called hard wiring logic or hard wired.

The same connections are made in the programmable controller by means of the programming commands. Although the programmable controller does not internally hard wire the individual components, but performs the same logic through the electrical circuit, the principle is the same. Following is a description of each basic instruction.

1. LD COMMAND
A logical string always begins with an LD (load) command. This command states that you are now starting a new string in the control circuit. The LD (load) command must be followed by an instruction telling which input is starting the string concerned. For example, if the string starts with input 5, the instruction would be LD IN 5 for an open contact, or LD NOT IN 5 for a closed contact.

2. AND COMMAND
This command indicates a connection in series between the last result and the next input.

3. OR COMMAND
This command is used to indicate a parallel connection between the last result and the next input.

4. = COMMAND
This command is used to actuate an output in accordance with the result of the logical input combinations.

TIMING AND COUNTING FUNCTIONS

Programmable controllers contain internal timers and counters which can be used anywhere in the program. The drawing below illustrates a typical line diagram using a timer and the required program to enter this logic into the programmable controller. In this circuit the output would be energized 5 seconds after input 1 and 2 were activated.

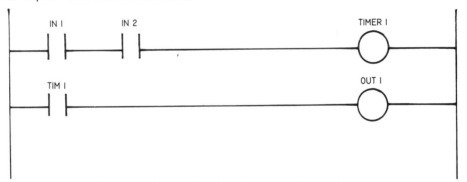

Memory location	INSTRUCTION		OPERAND		Title: date:
0	LD		IN	1	
1	AND		IN	2	
2	=		TIM	1	
3	TSET			50	5.0 SECONDS
4	LD		TIM	1	
5	=		OUT	1	
6	END				
7					
8					
9					

When using timers, the command immediately after = TIM (xxx) must state time delay. Most timers are programmed with a 0.1 graduation, so a time set of 55 would equal 5.5 seconds.

The drawing below illustrates a typical line diagram using a counter and the required program to enter this logic into the programmable controller. In this circuit the output would be energized after 15 inputs at input number 1.

Memory location	INSTRUCTION		OPERAND		Title: date:
0	LD		IN	1	
1	=		CNT	1	COUNT INPUTS
2	LD		IN	2	
3	=		CNTR	1	RESET COUNTER
4	LD		CNT	1	
5	=		OUT	1	
6	END				
7					
8					
9					

A counter has two inputs – a CNT (count) input for input of counting pulses, and a CNTR (count reset) input for reset of the counter. The required number of counting pulses must be programmed immediately after the CNTR command.

AND/OR COMBINATIONS

Special care must be taken when programming the OR command. This is because the OR command connects the entire previous result in parallel with the next input. For example, the following circuit is not directly programmable. This is because the command "OR OUT 1" would connect the OUT 1 contact all the way back to the power line, as illustrated in the next circuit.

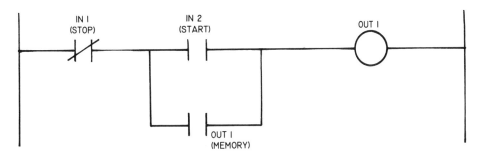

THIS CIRCUIT CANNOT BE PROGRAMMED AS IS.

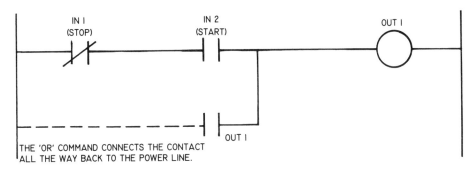

THE 'OR' COMMAND CONNECTS THE CONTACT
ALL THE WAY BACK TO THE POWER LINE.

This problem can be taken care of by rearranging the circuit as illustrated below and entering the following program.

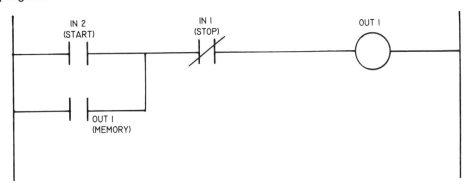

Memory location	INSTRUCTION		OPERAND		Title:	date:
0	LD		IN	2		
1	OR		OUT	1		
2	AND	NOT	IN	1		
3	=		OUT	1		
4	END					
5						
6						
7						
8						
9						

SY 115
Knob-adjustable

DATA SHEET M
SY 115

* **Load guard for asynchronous motors and other symmetrical loads.**
* **Measures phase difference (cos φ) between motor current and -voltage.**
* **Measuring range: Cos φ = 0 - 0.9 with current metering transformer type MI 100/MI 500.**
* **Knob-adjustable.**
* **With delayed function on start.**
* **10 A SPDT output relay.**
* **LED-indication for relay on.**
* **AC supply voltage.**

Courtesy of Electromatic Controls Corp.
2495 Pembroke Ave.
Hoffman Estates, Illinois

SPECIFICATIONS

Common technical data and ordering key
Pages 4 and 5. (Catalogue S 6.)

Supply voltage
3×380 VAC (type SY 115 380).
3×220 VAC (type SY 115 220).
Other voltages upon request.

Measuring range
Cos φ = 0 - 0.9.
With current metering transformer type MI 100 or MI 500.

Adjustment
Knob-adjustable with absolute scale (cos φ).

Hysteresis
10° equalling app. 1 graduation mark.

Start
On connection of supply voltage, the relay is energized for app. 5 seconds (T).

Measuring of current phase
Measuring input for connection of current metering transformer: Pins 8-11.

Voltage from current metering transformer:
0.1 - 4 V$_{peak}$.
If the current is below 2.5 A, the

conductor may be drawn through the central hole of the current metering transformer many times, so that the number of turns multiplied by the current consumption is inside the current range of the transformer.
The current metering transformer should be mounted in such a way that the current »flows« from the front towards the rear of the transformer.

Measuring
The voltage as well as the current are measured on the phase connected to pin 5.

Inversion
The output signal can be inverted by interconnecting pins 9 and 11.

Reaction time
During operation:
Typically 0.5 seconds.

Accessories
Bases.
Hold down spring.
Mounting rack.
Base covers.
Front mounting bezel.

Current metering transformer type MI 100 and MI 500.

WIRING DIAGRAMS

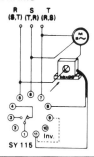

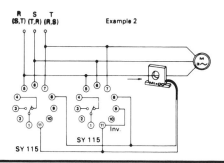

MODE OF OPERATION

This S-system can be used for monitoring the actual load of asynchronous motors. The SY 115 measures the angle between motor current and motor voltage, i.e. phase angle difference. This angle always exists and its change is almost proportional to the actual motor load (contrary to the motor current solely).
The characteristics of the load depend on the type of motor, and the phase difference, cos φ, depends upon the actual load. It is therefore recommended to adjust cos φ after practical tests.
The relay contact in the SY 115 should

be employed as a stop function in a system with external restart.

Example 1
The SY 115 is connected to a current metering transformer type MI as well as to a 3-phased asynchronous motor. The relay operates when cos φ is below the set value. At inversion (stippled line) the relay operates when cos φ exceeds the set value.

Example 2
By a combination of normal and inverted function, the SY 115 monitors whether cos φ is within a set maximum and minimum level respectively.

Phase difference/Load

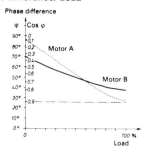

OPERATION DIAGRAM

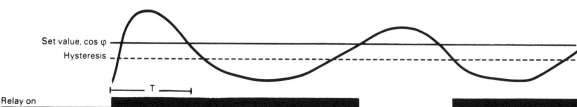

SY 185
Knob-adjustable

SY 185

❋ Relay for phase angle errors and phase breaking.
❋ Metering range for phase angle error: 5 - 15°.
❋ Knob-adjustable phase angle sensitivity.
❋ Operates irrespective of the phase sequence.
❋ 10 A SPDT output relay.
❋ LED-indication for relay on.
❋ Supply voltage is the 3-phased metering voltage.

Courtesy of Electromatic Controls Corp.
2495 Pembroke Ave.
Hoffman Estates, Illinois

SPECIFICATIONS

Inputs
Pins 5–6 and 7.

Metering-/supply voltage
3 x 220 VAC ⎫ ± 10 %.
3 x 380 VAC ⎭
Any phase sequence.

Frequency
50 Hz or 60 Hz.

Phase angle sensitivity
5 – 15° ± 10 %.
Knob-adjustable.

Amplitude sensitivity
± 30 %.

Hysteresis
Approx. 2°.

Reaction time on phase angle error
1 s.
Is available upon request with reaction times up to app. 4 s.

Accessories
Bases.
Hold down spring.
Mounting rack.
Base cover.
Front mounting bezel.

WIRING DIAGRAM

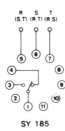

SY 185

MODE OF OPERATION

The relay meters on its own 3-phase supply voltage and controls the mutual phase angle.
It operates, irrespective of the phase sequence, when the angle error is smaller than the set value.

In case of interruption of a phase the relay will release, provided that the mutual phase angle error between the flawless phases and the phase possibly regenerated by electric motors connected, exceeds the set value.

Even if the phase angle error does not exceed the set value the relay shall release in case of phase breaking, provided that the voltage regenerated is below 70 % of the nominal voltage.

OPERATION DIAGRAMS

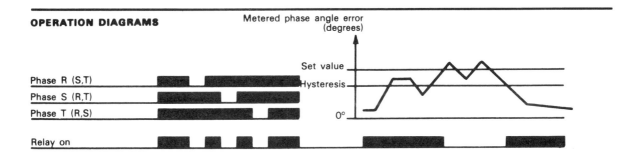

DATA SHEET O

SM 170
No
adjustment

SM 170

❋ **Phase sequence/Phase breaking relay.**
❋ **Measures on voltage.**
❋ **10 A SPDT output relay.**
❋ **LED-indication for relay on.**
❋ **Supply voltage is the 3-phased measuring voltage.**

Courtesy of Electromatic Controls Corp.
2495 Pembroke Ave.
Hoffman Estates, Illinois

SPECIFICATIONS

Inputs
Pins 5 - 6 and 7.

**Measuring voltage -
Supply voltage**
3 x 220 VAC ⎫ ± 10 %.
3 x 380 VAC ⎭

The measuring voltage also
works as
supply voltage.

Frequency
45 - 65 Hz.

Accessories
Bases.
Hold down spring.
Mounting rack.
Base cover.
Front mounting bezel.

WIRING DIAGRAMS

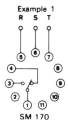

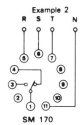

MODE OF OPERATION

The relay measures on its own 3-phased
supply voltage and operates, when all
phases are present and the phase sequence is correct.

Example 1
The relay releases in case of interruption
of one of the phases, provided that the
voltage regenerated by electric motors
on the interrupted phase does not exceed 50 % of the nominal voltage.

Example 2
If the value of the regenerated voltage is
somewhat higher than the 50 %
mentioned in example 1, the S-systen
can be brought to release, when the zero
line of the supply voltage is connected
to pin 11, as the sensitivity of the system thereby is improved.
At regenerated voltages the amplitude
depends on the size and the load of the
connected electric motor.
In practice the value of the regenerated
voltage can be near the same as the
value of the supply voltage.

OPERATION DIAGRAM

Phase »R« Pin 5				S	T	R	
Phase »S« Pin 6				R	S	S	
Phase »T« Pin 7				T	R	T	
Relay on							

238

SM 115
Knob-adjustable

SM 115

❋ **Current metering relay for AC.**
❋ **Metering range: 0.1-500 A (peak) with current metering transformer.**
❋ **Knob-adjustable trip point.**
❋ **Latching at set level possible.**
❋ **10 A SPDT output relay.**
❋ **LED-indication for relay on.**
❋ **AC- or DC supply voltage.**

Courtesy of Electromatic Controls Corp.
2495 Pembroke Ave.
Hoffman Estates, Illinois

SPECIFICATIONS

Input voltage
Pins 5-7: 0.1-4 V.
Max. 50 V.
Pin 5 positive.

Hysteresis
App. 10 %.
The hysteresis can be extended to app. 75 % by connecting a resistor between pins 8-9. Resistor limits are 1 MΩ and 15 KΩ. The hysteresis increases by decreasing resistance.

Latching
The relay shall latch at set level when pins 8 - 9 are interconnected.

**AC measurements
1- or 3-phases**
Are made in conjunction with one of the current metering transformers type MI or type MP. These transformers deliver an output voltage between 0.1 and 4 V being proportional with the current flowing in a conductor, drawn through the central hole of the transformer.

Accessories
Bases.
Hold down spring.
Mounting rack.
Base cover.
Front mounting bezel.
Current metering transformers type MI and type MP.

WIRING DIAGRAMS

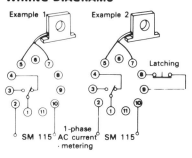

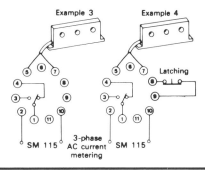

MODE OF OPERATION

Example 1
AC CURRENT METERING (1-phase). The relay operates when the current reaches the set value. The relay releases when the current drops at least 10 % below the set value (see hysteresis) or by disconnecting the supply voltage.

Example 2
AC CURRENT METERING (1-phase). LATCHING.
The relay operates when the current reaches the set value and then latches in operating position. The relay releases by removing the latch, i. e. by opening the

contact between pins 8-9, provided that the current has dropped at least 10 % below the set value (see hysteresis), or by disconnecting the supply voltage.

Example 3
AC CURRENT METERING (3 phases). The relay operates when the current in any of the phases reaches the set value. The relay releases when the current in all 3 phases drops at least 10 % below the set value (see hysteresis) or by disconnecting the supply voltage.

Example 4
AC CURRENT METERING (3 phases). LATCHING.
The relay operates when the current in any of the phases reaches the set value and then latches in the operating position. The relay releases by removing the latch, i. e. by opening the contact between pins 8-9, provided that the current in all 3 phases has dropped at least 10 % below the set value (see hysteresis), or by disconnecting the supply voltage.

OPERATION DIAGRAMS

Example 1-3

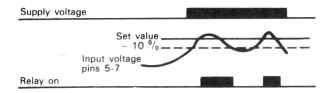

Example 2-4

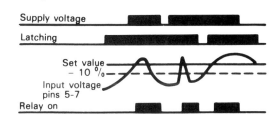

Standard Industrial Electrical Systems

DISCONNECT	CIRCUIT INTERRUPTER	CIRCUIT BREAKER W/THERMAL O.L.	CIRCUIT BREAKER W/MAGNETIC O.L.	CIRCUIT BREAKER W/THERMAL AND MAGNETIC O.L.

LIMIT SWITCHES		FOOT SWITCHES	PRESSURE & VACUUM SWITCHES	LIQUID LEVEL SWITCH	TEMPERATURE ACTUATED SWITCH	FLOW SWITCH (AIR, WATER, ETC.)
NORMALLY OPEN	NORMALLY CLOSED					

| HELD CLOSED | HELD OPEN | N.C. | N.C. | N.C. | N.C. | N.C. |

SPEED (PLUGGING)		ANTI-PLUG	SYMBOLS FOR STATIC SWITCHING CONTROL DEVICES

STATIC SWITCHING CONTROL IS A METHOD OF SWITCHING ELECTRICAL CIRCUITS WITHOUT THE USE OF CONTACTS, PRIMARILY BY SOLID STATE DEVICES. USE THE SYMBOLS SHOWN IN TABLE EXCEPT ENCLOSED IN DIAMOND:

EXAMPLES —
INPUT "COIL"   OUTPUT N.O.   LIMIT SW. N.O.   LIMIT SW. N.C.

SELECTOR

2 POSITION	3 POSITION	2 POS. SEL. PUSH BUTTON

CONTACTS	SELECTOR POSITION			
	A		B	
	BUTTON		BUTTON	
	FREE	DEPRES'D	FREE	DEPRES'D
1 - 2	X			
3 - 4		X	X	X

X-CONTACT CLOSED   X-CONTACT CLOSED   X-CONTACT CLOSED

PUSH BUTTONS

MOMENTARY CONTACT				MAINTAINED CONTACT		ILLUMINATED
SINGLE CIRCUIT	DOUBLE CIRCUIT	MUSHROOM HEAD	WOBBLE STICK	TWO SINGLE CIRCUIT	ONE DOUBLE CIRCUIT	
	N.O. & N.C.					

Standard Industrial Electrical Systems

CONTACTS								OVERLOAD RELAYS	
INSTANT OPERATING				TIMED CONTACTS – CONTACT ACTION RETARDED AFTER COIL IS:				THERMAL	MAGNETIC
WITH BLOWOUT		WITHOUT BLOWOUT		ENERGIZED		DE-ENERGIZED			
N.O.	N.C.	N.O.	N.C.	N.O.T.C.	N.C.T.O.	N.O.T.O.	N.C.T.C.		

SUPPLEMENTARY CONTACT SYMBOLS						
SPST N.O.		SPST N.C.		SPDT		TERMS
SINGLE BREAK	DOUBLE BREAK	SINGLE BREAK	DOUBLE BREAK	SINGLE BREAK	DOUBLE BREAK	SPST – SINGLE POLE SINGLE THROW
						SPDT – SINGLE POLE DOUBLE THROW
DPST, 2 N.O		DPST, 2 N.C.		DPDT		DPST – DOUBLE POLE SINGLE THROW
SINGLE BREAK	DOUBLE BREAK	SINGLE BREAK	DOUBLE BREAK	SINGLE BREAK	DOUBLE BREAK	DPDT – DOUBLE POLE DOUBLE THROW
						N.O. – NORMALLY OPEN
						N.C. – NORMALLY CLOSED

METER (INSTRUMENT)		PILOT LIGHTS	
INDICATE TYPE BY LETTER	TO INDICATE THE FUNCTION OF THE METER OR INSTRUMENT, PLACE THE SPECIFIED LETTER OR LETTERS WITHIN THE SYMBOL.	INDICATE COLOR BY LETTER	
		NON PUSH-TO-TEST	PUSH-TO-TEST
	AM – AMMETER VA – VOLT AMMETER AH – AMPERE-HOUR VAR – VARMETER μA – MICROAMMETER VARH – VARHOUR METER mA – MILLAMMETER W – WATTMETER PF – POWER FACTOR WH – WATTHOUR METER V – VOLTMETER		

INDUCTORS	COILS			
IRON CORE	DUAL VOLTAGE MAGNET COILS			BLOWOUT COIL
	HIGH VOLTAGE	LOW VOLTAGE		
AIR CORE				

Standard Industrial Electrical Systems

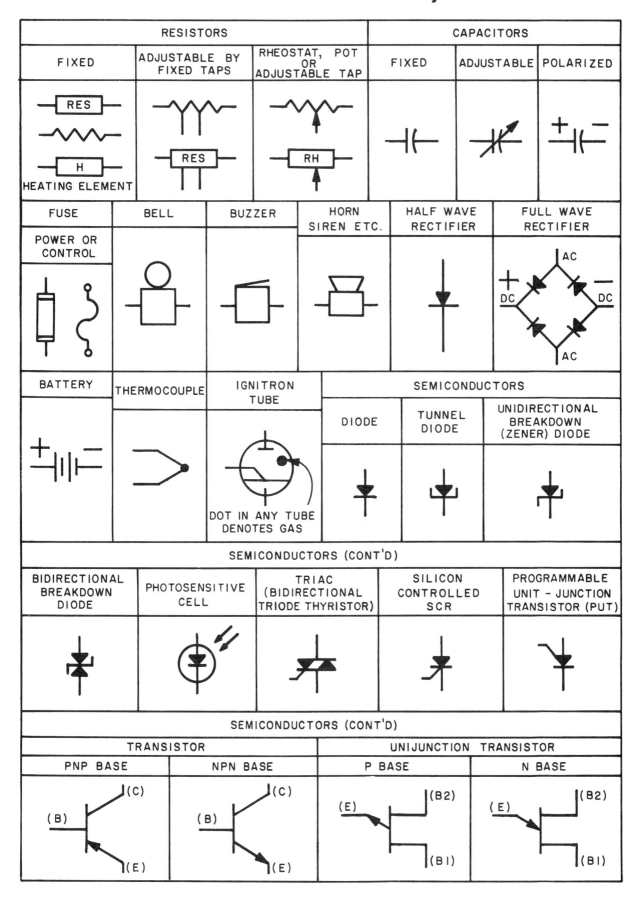

Standard Industrial Electrical Systems

TRANSFORMERS

AUTO	AIR CORE	CURRENT	CONTROL TRANSFORMER		AUTO-TRANSFORMER FOR REDUCED VOLTAGE STARTING
			SINGLE VOLTAGE	DUAL VOLTAGE	

Single Voltage: H1, H2, X2, X1
Dual Voltage: H1, H3, H2, H4, X2, X1

Auto-transformer for reduced voltage starting percentages: 50, 65, 80, 100, 0 (both sides)

AC MOTORS

SINGLE PHASE	SINGLE PHASE TWO-SPEED	THREE PHASE	SEPARATE WINDING TWO-SPEED	CONSTANT TORQUE TWO-SPEED

SINGLE PHASE: T1, T2
SINGLE PHASE TWO-SPEED: HIGH COM LOW — T1, T2, T3
THREE PHASE: T1, T2, T3
SEPARATE WINDING TWO-SPEED: T1, T3, T2 / T11, T13, T12
CONSTANT TORQUE TWO-SPEED: T4, T3, T1, T5, T2, T6

VARIABLE TORQUE TWO-SPEED	CONSTANT HORSEPOWER TWO-SPEED	WYE-DELTA REDUCED VOLTAGE	WYE CONNECTED PART WINDING REDUCED VOLTAGE

VARIABLE TORQUE TWO-SPEED: T4, T3, T1, T5, T2, T6
CONSTANT HORSEPOWER TWO-SPEED: T4, T3, T1, T5, T2, T6
WYE-DELTA REDUCED VOLTAGE: T6, T1, T3, T4, T5, T2
WYE CONNECTED PART WINDING REDUCED VOLTAGE: T1, T2, T3, T5, T7, T8, T9, T4, T6

DC MOTORS | WIRING | CONNECTIONS

ARMATURE	SHUNT FIELD	SERIES FIELD	COMM. OR COMPENS. FIELD	NOT CONNECTED	POWER	WIRING TERMINAL	MECHANICAL
ARM	(SHOW 4 LOOPS)	(SHOW 3 LOOPS)	(SHOW 2 LOOPS)	CONNECTED	CONTROL	GROUND	MECHANICAL INTERLOCK

CONTROL AND POWER CONNECTIONS - 600 VOLTS OR LESS - ACROSS-THE-LINE STARTERS

	1 PHASE	2 PHASE 4 WIRE	3 PHASE
LINE MARKINGS	L1, L2	L1, L3-PHASE 1 / L2, L4-PHASE 2	L1, L2, L3
GROUND WHEN USED	L1 IS ALWAYS UNGROUNDED	—	L2
MOTOR RUNNING OVERCURRENT UNITS IN — 1 ELEMENT	L1	—	—
2 ELEMENT	—	L1, L4	—
3 ELEMENT	—		L1, L2, L3
CONTROL CIRCUIT CONNECTED TO	L1, L2	L1, L3	L1, L2
FOR REVERSING INTERCHANGE LINES	—	L1, L3	L1, L3

Three-Phase 230 Volt Motors and Circuits. (240V System)

Size of Motor (Table 430-150) HP	Amp Rating	Motor Overload Protection — LOW-PEAK or FUSETRON Fuses (Amps): Motor Rated Not Over 40°C Or Not Less Than 1.15 S.F. (Max. Fuse 125%)	All Other Motors (Max. Fuse 115%)	Back-up Motor Protection: Motor Rated Not Over 40°C Or Not Less Than 1.15 S.F.	All Other Motors	Switch (115% Min. Or HP Rated) Or Fuseholder Size	Min. Size of Starter (NEMA)	60°C TW	60°C THW	60°C THWN	60°C THHN	60°C XHHW§	75°C TW	75°C THW	75°C THWN	75°C THHN	75°C XHHW§	Wire Size (AWG or kcmil)	Conduit (inches)
½	2	2½	2¼	2½	2½	30	00	●	●	●	●	●	●	●	●	●	●	14	½
¾	2.8	3½	3 2/10	3½	3½	30	00	●	●	●	●	●	●	●	●	●	●	14	½
1	3.6	4½	4	4½	4½	30	00	●	●	●	●	●	●	●	●	●	●	14	½
1½	5.2	6¼	5 6/10	7	6	30	00	●	●	●	●	●	●	●	●	●	●	14	½
2	6.8	8	7½	9	8	30	0	●	●	●	●	●	●	●	●	●	●	14	½
3	9.6	12	10	12	12	30	0	●	●	●	●	●	●	●	●	●	●	14	½
5	15.2	17½	17½	20	17.5	30	1	●	●	●	●	●	●	●	●	●	●	14	½
7½	22	25	25	30	30	30	1	●	●	●	●	●	●	●	●	●	●	10	½
10	28	35	30*	35	35	60	2	●	●				●					8	¾
										●	●	●						8	½
														●	●	●	●	10	½
15	42	50	45	60	50	60	2	●	●				●	●				6	1
										●	●	●			●	●	●	6	¾
20	54	60*	60*	70	70	100	3	●	●	●	●	●	●	●	●	●	●	4	1
25	68	80	75	90	80	100	3	●	●				●					3	1¼
										●	●	●						3	1
														●	●	●	●	4	1
30	80	100	90	100	100	100	3	●	●	●	●	●	●					1	1¼
														●				3	1¼
															●	●	●	3	1
40	104	125	110	150	125	200	4	●	●	●	●	●	●					2/0	1½
														●	●	●	●	1	1¼
50	130	150	150	175	150	200	4	●	●				●					3/0	2
										●	●	●						3/0	1½
														●	●	●	●	2/0	1½
60	154	175	175	200	200	200	5	●	●	●	●	●	●					4/0	2
														●				3/0	2
															●	●	●	3/0	1½
75	192	225	200*	250	225	400	5	●	●				●					300	2½
										●	●	●						300	2
														●				250	2½
															●	●	●	250	2
100	248	300	250	350	300	400	5	●	●	●	●	●	●					500	3
														●	●	●	●	350	2½
125	312	350	350	400	400	400	6	●	●	●	●	●	●					4/0-2/φ■	2-2■
														●				3/0-2/φ■	2-2■
															●	●	●	3/0-2/φ■	2-1½■
150	360	450	400*	450	450	600	6	●	●				●					300-2/φ■	2-2½■
										●	●	●						300-2/φ■	2-2■
														●	●	●	●	4/0-2/φ■	2-2■
200	480	600	500	600	600	600	6	●	●	●	●	●	●					500-2/φ■	2-3■
														●	●	●	●	350-2/φ■	2-2½■

■Indicates two sets of multiple conductors and two runs of conduit.
††Consult equipment manufacturer for their equipment's U.L. listed termination temperature.
*Fuse reducers required.
§For dry locations only.

Bussmann Div. Cooper Industries

Single-Phase Motors and Circuits.

1		2		3		4	5	6										7	
Size of Motor (Table 430-148)		**Motor Overload Protection**		**Back-up Motor Protection**		**Switch** (115% Min. Or HP Rated) **Or Fuse-holder Size**	**Min. Size of Starter** (NEMA)	**Controller Termination Temperature Rating**										**††Min. Size of Copper Wire and Trade Conduit**	
		LOW-PEAK or FUSETRON Fuses (Amps)						**60°C**					**75°C**						
HP	**Amp Rating**	Motor Rated Not Over 40°C Or Not Less Than 1.15 S.F. (Max. Fuse 125%)	All Other Motors (Max. Fuse 115%)	Motor Rated Not Over 40°C Or Not Less Than 1.15 S.F.	All Other Motors			**TW**	**THW**	**THWN**	**THHN**	**XHHW§**	**TW**	**THW**	**THWN**	**THHN**	**XHHW§**	**Wire Size** (AWG or kcmil)	**Con-duit** (inches)
115 VOLTS (120V System)**																			
⅙	**4.4**	**5**	**5**	**5 6/10**	**5 6/10**	**30**	00	■	■	■	■	■	■	■	■	■	■	14	½
¼	**5.8**	**7**	**6¼**	**7½**	**7**	**30**	00	■	■	■	■	■	■	■	■	■	■	14	½
⅓	**7.2**	**9**	**8**	**9**	**9**	**30**	0	■	■	■	■	■	■	■	■	■	■	14	½
½	**9.8**	**12**	**10**	**15**	**12**	**30**	0	■	■	■	■	■	■	■	■	■	■	14	½
¾	**13.8**	**15**	**15**	**17½**	**17½**	**30**	0	■	■	■	■	■	■	■	■	■	■	14	½
1	**16**	**20**	**17½**	**20**	**20**	**30**	0	■	■	■	■	■	■	■	■	■	■	14	½
1½	**20**	**25**	**20**	**25**	**25**	**30**	1	■	■	■	■	■	■	■	■	■	■	12	½
2	**24**	**30**	**25**	**30**	**30**	**30**	1	■	■	■	■	■	■	■	■	■	■	10	½
230 VOLTS (240V System)																			
⅙	**2.2**	**2½**	**2½**	**2 8/10**	**2 8/10**	**30**	00	■	■	■	■	■	■	■	■	■	■	14	½
¼	**2.9**	**3½**	**3 2/10**	**4**	**3½**	**30**	00	■	■	■	■	■	■	■	■	■	■	14	½
⅓	**3.6**	**4½**	**4**	**4½**	**4½**	**30**	00	■	■	■	■	■	■	■	■	■	■	14	½
½	**4.9**	**5 6/10**	**5 6/10**	**6¼**	**6**	**30**	00	■	■	■	■	■	■	■	■	■	■	14	½
¾	**6.9**	**8**	**7½**	**9**	**8**	**30**	00	■	■	■	■	■	■	■	■	■	■	14	½
1	**8**	**10**	**9**	**10**	**10**	**30**	00	■	■	■	■	■	■	■	■	■	■	14	½
1½	**10**	**12**	**10**	**15**	**12**	**30**	0	■	■	■	■	■	■	■	■	■	■	14	½
2	**12**	**15**	**12**	**15**	**15**	**30**	0	■	■	■	■	■	■	■	■	■	■	14	½
3	**17**	**20**	**17½**	**25**	**20**	**30**	1	■	■	■	■	■	■	■	■	■	■	12	½
									■									8	¾
5	**28**	**35**	**30***	**35**	**35**	**60**	2	■		■	■	■	■					8	½
														■	■	■	■	10	½
								■	■	■	■	■	■					6	¾
7½	**40**	**50**	**45**	**50**	**50**	**60**	2							■				8	¾
															■	■	■	8	½
								■	■				■					4	1
10	**50**	**60**	**50**	**70■**	**60**	**60**	3			■	■	■						4	¾
														■	■	■	■	6	¾

**On 125 Volts or less, Fustat Type S fuses or Fusetron dual-element plug fuses can be used in place of 0 to 14 ampere Fusetron dual-element cartridge fuses.
§For dry locations only.
*Fuse reducers required.
††Consult equipment manufacturer for their equipment's U.L. listed termination temp.
■100A Switch required.

Bussmann Div. Cooper Industries

Three-Phase 460 Volt Motors and Circuits. (480V System)

1 Size of Motor (Table 430-150) HP	1 Amp Rating	2 Motor Overload Protection — LOW-PEAK or FUSETRON Fuses (Amps): Motor Rated Not Over 40°C Or Not Less Than 1.15 S.F. (Max. Fuse 125%)	2 Motor Overload Protection: All Other Motors (Max. Fuse 115%)	3 Back-up Motor Protection: Motor Rated Not Over 40°C Or Not Less Than 1.15 S.F.	3 Back-up Motor Protection: All Other Motors	4 Switch (115% Min. Or HP Rated) Or Fuse-holder Size	5 Min. Size of Starter (NEMA)	6 Controller Termination Temperature Rating 60°C TW	60°C THW	60°C THWN	60°C THHN	60°C XHHW§	75°C TW	75°C THW	75°C THWN	75°C THHN	75°C XHHW§	7 ††Min. Size of Copper Wire and Trade Conduit: Wire Size (AWG or kcmil)	Conduit (inches)
½	1	1¼	1⅛	1¼	1¼	30	00	■	■	■	■	■	■	■	■	■	■	14	½
¾	1.4	1 8/10	1 6/10	1 8/10	1 8/10	30	00	■	■	■	■	■	■	■	■	■	■	14	½
1	1.8	2¼	2	2¼	2¼	30	00	■	■	■	■	■	■	■	■	■	■	14	½
1½	2.6	3 2/10	2 8/10	3½	3	30	00	■	■	■	■	■	■	■	■	■	■	14	½
2	3.4	4	3½	4½	4	30	00	■	■	■	■	■	■	■	■	■	■	14	½
3	4.8	5 6/10	5	6	5 6/10	30	0	■	■	■	■	■	■	■	■	■	■	14	½
5	7.6	9	8	10	9	30	0	■	■	■	■	■	■	■	■	■	■	14	½
7½	11	12	12	15	15	30	1	■	■	■	■	■	■	■	■	■	■	14	½
10	14	17½	15	17½	17½	30	1	■	■	■	■	■	■	■	■	■	■	14	½
15	21	25	20	30	25	30	2	■	■	■	■	■	■	■	■	■	■	10	½
20	27	30*	30*	35	35	60	2	■	■			■	■					8	¾
										■	■							8	½
														■	■	■	■	10	½
25	34	40	35	45	40	60	2	■	■				■					6	1
										■	■	■						6	¾
														■			■	8	¾
															■	■		8	½
30	40	50	45	50	50	60	3	■	■				■					6	1
										■	■	■						6	¾
														■			■	8	¾
															■	■		8	½
40	52	60*	60*	70	60*	100	3	■	■	■	■	■	■					4	1
															■	■	■	6	¾
														■				6	1
50	65	80	70	90	75	100	3	■	■				■					3	1¼
										■	■	■						3	1
														■	■	■	■	4	1
60	77	90	80	100	90	100	4	■	■	■	■	■	■					1	1¼
															■	■	■	3	1
														■				3	1¼
75	96	110	110	125	125	200	4	■	■				■					1/0	1½
										■	■	■						1/0	1¼
														■	■	■	■	1	1¼
100	124	150	125	175	150	200	4	■	■				■					3/0	2
										■	■	■						3/0	1½
														■	■	■	■	2/0	1½
125	156	175	175	200	200	200	5	■	■	■	■	■	■					4/0	2
														■				3/0	2
															■	■	■	3/0	1½
150	180	225	200*	225	225	400	5	■	■				■					300	2½
										■	■	■						300	2
														■	■	■	■	4/0	2
200	240	300	250	300	300	400	5	■	■	■	■	■	■					500	3
														■	■	■	■	350	2½
250	302	350	325	400	350	400	6	■	■	■	■	■	■					4/0-2/ϕ■	2-2■
														■				3/0-2/ϕ■	2-2■
															■	■	■	3/0-2/ϕ■	2-1½■
300	361	450	400*	500	450	600	6	■	■				■					300-2/ϕ■	2-2½■
										■	■	■						300-2/ϕ■	2-2■
														■	■	■	■	4/0-2/ϕ■	2-2■

§For dry locations only.
††Consult equipment manufacturer for their equipment's U.L. listed termination temperature.
*Fuse reducers required.
■Indicates two sets of multiple conductors and two runs of conduit.

Bussmann Div. Cooper Industries

Direct Current Motors and Circuits.

1		2		3		4	5	6										7	
Size of Motor (Table 430-150)		Motor Overload Protection — Dual-Element Fuse		Back-up Motor Protection		Switch (115% Min. Or HP Rated) Or Fuseholder Size	Min. Size of Starter (NEMA)	Controller Termination Temperature Rating										††Min. Size of Copper Wire and Trade Conduit	
								60°C					75°C						
HP	Amp Rating	Motor Rated Not Over 40°C Or Not Less Than 1.15 S.F. (Max. Fuse 125%)	All Other Motors (Max. Fuse 115%)	Motor Rated Not Over 40°C Or Not Less Than 1.15 S.F.	All Other Motors			TW	THW	THWN	THHN	XHHW§	TW	THW	THWN	THHN	XHHW§	Wire Size (AWG or kcmil)	Conduit (inches)
90 VOLT																			
¼	4.0	5	4½	5	5	30	0	■	■	■	■	■	■	■	■	■	■	14	½
⅓	5.2	6¼	5⁶⁄₁₀	7	6	30	0	■	■	■	■	■	■	■	■	■	■	14	½
½	6.8	8	7½	9	8	30	0	■	■	■	■	■	■	■	■	■	■	14	½
¾	9.6	12	10	12	12	30	0	■	■	■	■	■	■	■	■	■	■	14	½
1	12.2	15	12	17½	15	30	0	■	■	■	■	■	■	■	■	■	■	14	½
120 VOLT																			
¼	3.1	3½	3½	4	4	30	0	■	■	■	■	■	■	■	■	■	■	14	½
⅓	4.1	5	4½	5⁶⁄₁₀	5	30	0	■	■	■	■	■	■	■	■	■	■	14	½
½	5.4	6¼	6	7	6¼	30	0	■	■	■	■	■	■	■	■	■	■	14	½
¾	7.6	9	8	10	9	30	0	■	■	■	■	■	■	■	■	■	■	14	½
1	9.5	10	10	12	12	30	0	■	■	■	■	■	■	■	■	■	■	14	½
1½	13.2	15	15	17½	17½	30	1	■	■	■	■	■	■	■	■	■	■	14	½
2	17	20	17½	25	20	30	1	■	■	■	■	■	■	■	■	■	■	12	½
3	25	30*	25*	35	30*	60	1	■		■	■	■	■					8	½
									■									8	¾
														■	■	■	■	10	½
5	40	50	45	50	50	60	2	■	■	■	■	■	■					6	¾
														■				8	¾
															■	■	■	8	½
7½	58	70	60*	75	70	100	3	■	■	■	■	■	■					3	1
														■				4	1
															■	■	■	4	¾
10	76	90	80	100	90	100	3	■	■	■	■	■	■					2	1
														■	■	■	■	3	1
180 VOLT																			
¼	2	2½	2¼	2½	2½	30	0	■	■	■	■	■	■	■	■	■	■	14	½
⅓	2.6	3²⁄₁₀	2⁸⁄₁₀	3½	3	30	0	■	■	■	■	■	■	■	■	■	■	14	½
½	3.4	4	3½	4½	4	30	0	■	■	■	■	■	■	■	■	■	■	14	½
¾	4.8	6	5	6	5⁶⁄₁₀	30	0	■	■	■	■	■	■	■	■	■	■	14	½
1	6.1	7½	7	8	7½	30	0	■	■	■	■	■	■	■	■	■	■	14	½
1½	8.3	10	9	12	10	30	1	■	■	■	■	■	■	■	■	■	■	14	½
2	10.8	12	12	15	15	30	1	■	■	■	■	■	■	■	■	■	■	14	½
3	16	20	17½	20	20	30	1	■	■	■	■	■	■	■	■	■	■	12	½
5	27	30*	30*	35	35	60	1	■		■	■	■	■					8	½
									■									8	¾
														■	■	■	■	10	½

†If manufacturer's overload relay table states a maximum branch circuit protective device of a lower rating, that lower maximum rating must be used in lieu of above recommendation (430-52).
§For dry locations only.
††Consult equipment manufacturer for their equipment's U.L. listed termination temperature.
*Fuse reducers required.

Bussmann Div. Cooper Industries

Common Abbreviations for Electrical Terms and Devices

AC	Alternating current		MB	Magnetic brake
ALM	Alarm		MCS	Motor circuit switch
AM	Ammeter		MEM	Memory
ARM	Armature		MTR	Motor
AU	Automatic		MN	Manual
BAT	Battery (electrical)		NEG	Negative
BR	Brake relay		NEUT	Neutral
CAP	Capacitor		NC	Normally closed
CB	Circuit breaker		NO	Normally open
CEMF	Counter electromotive force		OHM	Ohmmeter
CKT	Circuit		OL	Overload relay
CONT	Control		PB	Pushbutton
CR	Control relay		PH	Phase
CRM	Control relay master		PLS	Plugging switch
CT	Current transformer		POS	Positive
D	Down		PRI	Primary switch
DB	Dynamic braking contactor or relay		PS	Pressure switch
DC	Direct current		R	Reverse
DIO	Diode		REC	Rectifier
DISC	Disconnect switch		RES	Resistor
DP	Double pole		RH	Rheostat
DPDT	Double pole, double throw		S	Switch
DPST	Double pole, single throw		SCR	Semiconductor-controlled rectifier
DS	Drum switch		SEC	Secondary
DT	Double throw		1PH	Single phase
EMF	Electromotive force		SOC	Socket
F	Forward		SOL	Solenoid
FLS	Flow switch		SP	Single pole
FREQ	Frequency		SPDT	Single pole, double throw
FS	Float switch		SPST	Single pole, single throw
FTS	Foot switch		SS	Selector switch
FU	Fuse		SSW	Safety switch
GEN	Generator		T	Transformer
GRD	Ground		TB	Terminal board
IC	Integrated circuit		3PH	Three phase
INTLK	Interlock		TD	Time delay
IOL	Instantaneous overload		THS	Thermostat switch
JB	Junction box		TR	Time delay relay
LS	Limit switch		U	Up
LT	Lamp		UCL	Unclamp
M	Motor starter		UV	Under voltage